Banking and Financial Markets

Paola Ferretti · Pierluigi Martino

Banking and Financial Markets

New Risks and Challenges from Fintech and Sustainable Finance

palgrave
macmillan

Paola Ferretti
Department of Economics
and Management
University of Pisa
Pisa, Italy

Pierluigi Martino
Department of Economics
and Management
University of Pisa
Pisa, Italy

ISBN 978-3-031-32561-8 ISBN 978-3-031-32562-5 (eBook)
https://doi.org/10.1007/978-3-031-32562-5

Cover illustration: Melisa Hasan

This Palgrave Macmillan imprint is published by the registered company Springer Nature Switzerland AG
The registered company address is: Gewerbestrasse 11, 6330 Cham, Switzerland

Contents

List of Figures

CHAPTER 1

Introduction

Abstract This chapter sets out the objectives of the book and describes its main structure. In particular, it provides a general overview on the specific context of the book, and it presents a summary of the content of each chapter.

Keywords Financial markets · Financial intermediaries · Banking · FinTech · Sustainable finance

1.1 Introduction

This book aims to provide readers the foundation knowledge, essential for comprehending the basics of the functioning of the financial market and institutions, and the main challenges they currently face. Thereby, the book is particularly tailored to students, as well as anyone interested, who are encountering the financial and banking issues for the first time.

The financial system consists of the financial infrastructure and all financial intermediaries and financial markets, and their relations with respect to the flow of funds to and from households, governments, companies and other economic actors (de Haan et al., 2020). Its primary function is to facilitate the transfer of financial resources from savers (i.e. surplus units) to those who need funds (i.e. deficit units), thereby

P. Ferretti and P. Martino, *Banking and Financial Markets*,
https://doi.org/10.1007/978-3-031-32562-5_1

favouring a better allocation of resources and, consequently, contributing to economic growth (Boot & Thakor, 1997; Levine, 1997). Hence, a well-functioning financial system is a crucial prerequisite for economic development, and a large body of research provides empirical evidence to this argument showing a positive link between the development of the financial system and economic growth (Asteriou & Spanos, 2019; Hassan et al., 2011). This book examines how financial markets and financial intermediaries work and contribute to economic growth, describing the structure and the operational characteristics of financial markets, as well as the functions and the main business of financial intermediaries, with a particular focus on banks.

The book also outlines the most important steps in the European integration process, both on the monetary side and on the financial side. The evolutionary perspective helps to understand the decisions taken today, especially in the light of certain adverse events that occurred in the last decades. The subprime mortgage financial crisis and the subsequent sovereign debt crisis have severely undermined bank stability in the European Union (EU), raising macro-economic concerns among the various EU institutions. Notably, EU's banking system has shown itself to be vulnerable in a time of macro-economic and financial shocks owing to the fragmentation of its regulatory system and the weakness of its supervision system (Bremus & Lambert, 2014; Degl'Innocenti et al., 2017; Lane, 2012). In response to the financial and sovereign debt crises, EU authorities agreed to create an integrated financial framework, the European Banking Union (EBU), to take several coordinated steps to address flaws in the EU's banking sector, such as banks' low profitability and high non-performing loan (NPL) stocks, with the aim to reduce banks' risks and thus ensure the EU's financial stability (Colombini, 2015; Martino, 2019a). The origins of the international financial crisis and its impacts on the banking business are discussed, in addition to the actions adopted by regulators and supervisors. Moreover, the implications connected with the pandemic crisis we are still experiencing are addressed.

The structure of the banking sector is examined also illustrating the most important banking business areas: commercial versus investment banking. This has become a key topic in banking literature (Ayadi et al., 2016; Cosma et al., 2017). Commercial banks are typically smaller and provide lending and deposit-taking services for households and small and medium-sized enterprises. By contrast, investment banks fund most of their activities on wholesale markets and generate a large share of their income from non-traditional activities, such as investment banking and trading. The literature shows that such differences may affect banks risk and return, and consequently, banks stability (Altunbas et al., 2011; Beck et al., 2013; Demirgüc-Kunt & Huizinga, 2010; Kohler, 2015). To this end, the origins of such distinction and the main activities typically associated to each business area are analysed.

Lastly, the book introduces two important phenomena which are currently characterizing the financial environment.

Firstly, recent technological developments have led to the rise of financial technology (FinTech) sector, which covers digital innovations and technology-enabled business model innovations in the financial arena (Frame et al., 2018; FSB, 2017; Philippon, 2016). The advent of new technologies, such as blockchain, cryptocurrencies, artificial intelligence (AI), etc., may create many benefits for consumers and organizations, including access to credit and access to a wider product range. Additionally, financial intermediaries may benefit from the adoption of such technologies, which may boost processes' efficiency and, therefore, improve the services intermediaries offer. At the same time, new technologies may also pose new risks to the financial system, which policymakers, regulators and supervisors should consider ensuring the financial stability, safety and soundness of financial institutions (Martino, 2019b, 2021).

Secondly, sustainability has gained centrality in today's financial landscape. This is due to the growing attention by policymakers and industrial decision-makers around the world towards the transition to a low-carbon, more resource-efficient and sustainable economy (Scordato et al., 2018; Stoycheva et al., 2018). On the one hand, most banks today are modifying their approach to meet the needs of a green and sustainable economy, redesigning their products and services or carrying out initiatives consistent with environmental and social objectives (e.g., Avrampou et al., 2019; Cosma et al., 2020; Galletta et al., 2021; Khan et al., 2021). However, there are also challenges to address, particularly with regard to incorporating sustainability within banks' decision-making processes and business functions, such as risk management. On the other hand, financial markets are undergoing a deep transformation as institutional investors are increasingly attracted to a host of new financial instruments, particularly in the bonds market, such as green bonds, sustainability-linked and social bonds. These form part of a broader ESG endeavour to rely on private and public sources to fund sustainable investments.

Both FinTech and sustainable finance represent very significant sources of opportunities for the market and the operators, even if the challenges and risks associated are also relevant. This underlines the need to understand their intensity and their potential impact on the functioning of the financial system considered as a whole.

The book is structured as follows. Chapter 2 provides a general overview on the functioning of the financial system, examining how financial markets and financial intermediaries work and contribute to economic growth. Notably, each type of financial market (i.e. monetary, bonds, stocks and derivatives) is described with a focus on the financial instruments that are traded and the main participants in the markets. The business and the main functions of financial intermediaries are also discussed. These are represented by insurances, investment intermediaries and banks.

Chapter 3 outlines the evolution of the European integration process, giving an overview of the most important regulatory steps related to the banking and financial system culminated in the creation of the Banking Union. A focus is also given to the international financial crisis, its origins and impacts from the perspective of the European banking system and to the role played by supervision aimed at ensuring the financial stability.

Chapter 4 illustrates some of the possible banking business areas in favour of firms. To this end, a focus on the traditional lending activity is given as well as on other financial services a bank may provide to firms. The underlying logic is to understand the extension of the offer of financial products, which may promote the enhancing of the banks-firms relationship along a perspective of an increasing integration between financial markets and intermediaries.

Chapter 5 addresses the phenomena of FinTech, providing a discussion about its main implications for banks and financial markets. In particular, this chapter examines the implications on the banking industry, highlighting the opportunities as well as the risks that FinTech poses to banks' business. Moreover, it explores the implications for financial markets with specific regard to the market for entrepreneurial finance, where many new players have entered the arena (Bellavitis et al., 2017; Block et al., 2018). Specifically, this chapter focuses on the most important innovations in the market for entrepreneurial finance, namely crowdfunding and Initial Coin Offerings (ICOs).

Finally, Chapter 6 discusses the opportunities and risks deriving from sustainable finance, examining their intensity and their potential impact on the banking industry and financial markets. In particular, it addresses the role of banking in the transitioning to a sustainable economy, outlining the main regulatory developments in the industry and their implications for banks' business. Moreover, it explores new developments in financial markets, where a host of new financial instruments, such as green bonds, sustainability-linked and social bonds, are available to fund sustainable and environmental investments.

References

Altunbas, Y., Manganelli, S., & Marques-Ibanez, D. (2011). *Bank risk during the financial crisis: Do business models matter?* ECB Working Paper No. 1394. https://papers.ssrn.com/sol3/papers.cfm?abstract_id=1945143

Asteriou, D., & Spanos, K. (2019). The relationship between financial development and economic growth during the recent crisis: Evidence from the EU. *Finance Research Letters, 28*, 238–245.

Avrampou, A., Skouloudis, A., Iliopoulos, G., & Khan, N. (2019). Advancing the sustainable development goals: Evidence from leading European banks. *Sustainable Development, 27*(4), 743–757.

Ayadi, R., De Groen, W., Sassi, I., Mathlouthi, W., Rey, H., & Aubry, O. (2016). *Banking business models monitor 2015 Europe*. https://papers.ssrn.com/sol3/papers.cfm?abstract_id=2784334

Beck, T., Demirgüç-Kunt, A., & Merrouche, O. (2013). Islamic vs. conventional banking: Business model, efficiency and stability. *Journal of Banking & finance, 37*(2), 433–447.

Bellavitis, C., Filatotchev, I., Kamuriwo, D. S., & Vanacker, T. (2017). Entrepreneurial finance: New frontiers of research and practice: Editorial for the special issue embracing entrepreneurial funding innovations. *Venture Capital, 19*(1–2), 1–16.

Block, J. H., Colombo, M. G., Cumming, D. J., & Vismara, S. (2018). New players in entrepreneurial finance and why they are there. *Small Business Economics, 50*, 239–250.

Boot, A. W., & Thakor, A. V. (1997). Financial system architecture. *The Review of Financial Studies, 10*(3), 693–733.

Bremus, F. M., & Lambert, C. (2014). Banking union and bank regulation: Banking sector stability in Europe. *DIW Economic Bulletin, 4*(9), 29–39.

Colombini, F. (2015). Risk, regulation, supervision and crises in the European Banking Union. *Law and Economics Yearly Review, 4*(2), 236–273.

Cosma, S., Ferretti, R., Gualandri, E., Landi, A., & Venturelli, V. (2017). The business model of banks: A review of the theoretical and empirical literature. *The Business of Banking: Models, Risk and Regulation*, 131–167.

Cosma, S., Venturelli, A., Schwizer, P., & Boscia, V. (2020). Sustainable development and European banks: A non-financial disclosure analysis. *Sustainability, 12*(15), 6146.

De Haan, J., Schoenmaker, D., & Wierts, P. (2020). *Financial markets and institutions: A European perspective*. Cambridge University Press.

Degl'Innocenti, M., Kourtzidis, S. A., Sevic, Z., & Tzeremes, N. G. (2017). Bank productivity growth and convergence in the European Union during the financial crisis. *Journal of Banking & Finance, 75*, 184–199.

Demirgüç-Kunt, A., & Huizinga, H. (2010). Bank activity and funding strategies: The impact on risk and returns. *Journal of Financial Economics, 98*(3), 626–650.

Frame, W. S., Wall, L. D., & White, L. J. (2018). *Technological change and financial innovation in banking: Some implications for FinTech*. Federal Reserve Bank of Atlantia, Working paper series. https://www.frbatlanta.org/-/media/documents/research/publications/wp/2018/11-technological-change-and-financial-innovation-in-banking-some-implications-for-fintech-2018-10-02.pdf

FSB. (2017, June 27). *Financial stability implications from FinTech: Supervisory and regulatory issues that merit authorities' attention*. https://www.fsb.org/wp-content/uploads/R270617.pdf

Galletta, S., Mazzù, S., Naciti, V., & Vermiglio, C. (2021). Sustainable development and financial institutions: Do banks' environmental policies influence customer deposits? *Business Strategy and the Environment, 30*(1), 643–656.

Hassan, M. K., Sanchez, B., & Yu, J. S. (2011). Financial development and economic growth: New evidence from panel data. *The Quarterly Review of Economics and Finance, 51*(1), 88–104.

Khan, H. Z., Bose, S., Mollik, A. T., & Harun, H. (2021). "Green washing" or "authentic effort"? An empirical investigation of the quality of sustainability reporting by banks. *Accounting, Auditing & Accountability Journal, 34*(2), 338–369.

Köhler, M. (2015). Which banks are more risky? The impact of business models on bank stability. *Journal of Financial Stability, 16*, 195–212.

Lane, P. R. (2012). The European sovereign debt crisis. *Journal of Economic Perspectives, 26*(3), 49–68.

Levine, R. (1997). Financial development and economic growth: Views and agenda. *Journal of Economic Literature, 35*(2), 688–726.

Martino, P. (2019a). *Non-performing loans in European banks: Management and resolution*. Franco Angeli, Milano.

Martino, P. (2019b). Blockchain technology: Challenges and opportunities for banks. *International Journal of Financial Innovation in Banking, 2*(4), 314–333.

Martino, P. (2021). *Blockchain and banking: How technological innovations are shaping the banking industry*. Springer Nature.

Philippon, T. (2016). *The fintech opportunity* (No. w22476). National Bureau of Economic Research working paper series. https://www.nber.org/system/files/working_papers/w22476/w22476.pdf

Scordato, L., Klitkou, A., Tartiu, V. E., & Coenen, L. (2018). Policy mixes for the sustainability transition of the pulp and paper industry in Sweden. *Journal of Cleaner Production, 183*, 1216–1227.

Stoycheva, S., Marchese, D., Paul, C., Padoan, S., Juhmani, A. S., & Linkov, I. (2018). Multi-criteria decision analysis framework for sustainable manufacturing in automotive industry. *Journal of Cleaner Production, 187*, 257–272.

CHAPTER 2

The Financial System: The European Union Perspective

Abstract This chapter provides a general overview on the role and functioning of the financial system, with a specific focus on the EU context. In particular, the chapter presents a discussion of the principal financial markets, namely money, bonds, equity and derivatives markets, highlighting their different characteristics in terms of market participants and their purposes, as well as in terms of features of the financial instruments negotiated. Moreover, it explores the fundamentals behind financial intermediation and provides a general overview on the business and main characteristics of the principal categories of financial intermediaries.

Keywords Financial system · Financial markets · Financial instruments · Financial intermediaries · European Union

2.1 The Role and Structure of the Financial System

The financial system performs the essential economic function of channelling funds from economic actors that have saved surplus funds, by spending less than their income, to those that have a shortage of funds because they wish to spend more than their income (Boot & Thakor, 1997; Merton & Bodie, 1995). Those participants who receive more

P. Ferretti and P. Martino, *Banking and Financial Markets*,
https://doi.org/10.1007/978-3-031-32562-5_2

money than they spend are referred to as *surplus units* (or *lender-savers*): they provide their net savings to the financial system in order to earn a return on their investments. Those participants who spend more money than they receive are referred to as *deficit units* (or *borrower-spenders*): they access funds from the financial system so that they can spend more money than they receive. In general, the principal lender-savers are households, even if it is worth noting that non-financial corporations and the government, as well as foreigners and their governments, also find themselves with excess funds and so lend them out. On the other hand, the most important borrower-spenders are generally non-financial corporations and the government, but households and foreigners also borrow financial resources to finance their expenditures.[1]

By channelling funds from surplus units to deficit units, the financial system is central in producing an efficient allocation of capital which contributes to higher production and efficiency for the overall economy (Beck et al., 2000; Xu et al., 2021). In particular, by allowing funds to move from actors who do not have a productive use to those who do, a well-functioning financial system has direct effects on personal wealth, the behaviour of businesses and consumers, and ultimately the cyclical performance of the economy. In other words, it enables families to obtain mortgages to finance their expenditures (e.g., purchases of cars, furniture, and houses), non-financial corporations to finance their growth, and governments to finance many of their expenditures (e.g., infrastructure construction like schools, hospitals and other projects), thereby contributing to the economic growth. The positive contribution of the financial system to economic growth is widely acknowledged in economic literature, with a large number of studies that provide empirical support to the argument that well-developed financial systems (financial markets and banking system) foster countries' economic growth (Beck et al., 2000; Beck & Levine, 2002; Levine et al., 2000; Levine & Zervos, 1998; Mishra & Narayan, 2015; Rajan & Zingales, 1996).

As shown in Fig. 2.1, the transferring of funds from lender-savers to borrower-spenders can occur via two channels, which identify two different structures of the financial system, i.e. *market-based* versus *banked-based* system.

[1] See, for more details, data from the latest Statistics bulletin by the European Central Bank on *Integrated economic and financial accounts by institutional sector* (last updated on 28-10-2022).

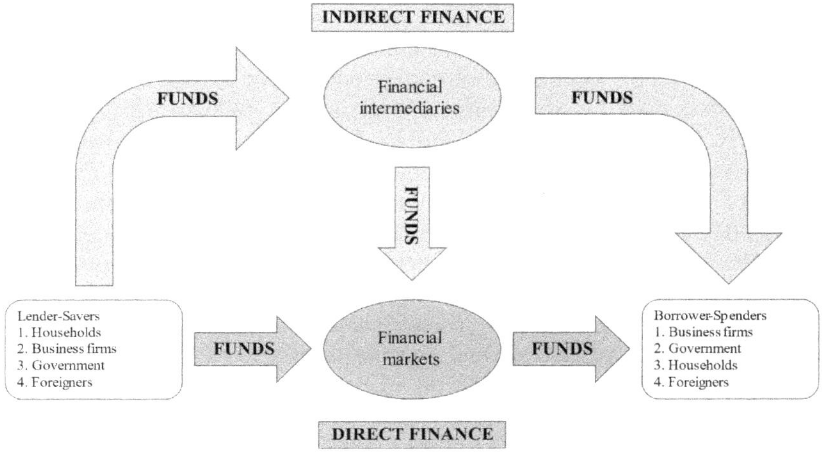

Fig. 2.1 The structure of the financial system (*Source* Mishkin and Eakins 2012)

In a market-based system (*direct financing* channel), deficit units borrow funds directly from lenders in financial markets by selling them *securities* (also called *financial instruments*), which are claims on the issuer's future income or assets (Mishkin & Eakins, 2012). Securities are assets for the actor who buys them, while they are liabilities for the actor that sells (issues) them. Here, non-financial corporations that need to borrow funds to finance their investments (e.g., purchase new factory and plants) may borrow the funds from savers by selling them financial instruments, such as *bonds* (i.e. a debt security that promises to make payments periodically for a specified period of time) or *stocks* (a security that entitles the owner to a share of the company's profits and assets). Hence, it is predominantly through financial markets that deficit units interact with those providing the capital. On the other hand, in a bank-based system (*indirect financing* channel), the process of channelling funds involves a financial intermediary like a bank that helps transfer funds from lender-savers to borrower-spenders. In particular, a financial intermediary does this by borrowing funds from lender-savers and then using these funds to make loans to borrower-spenders. As an example, a bank might acquire funds by issuing a liability to the public in the form of savings deposits, and it might then use the funds to acquire an asset by making a loan to companies and households. In this system, banks are crucial players in channelling funds from investors to deficit units (Thakor & Boot, 2008).

The relative development of direct versus indirect finance varies considerably across countries, with some relying more on financial markets (e.g., the USA), while others relying more on banks (e.g., EU member states, such as Austria, Hungary, Italy and Germany, among others) as the main source of financing (Demirgüç-Kunt & Maksimovic, 2002; Gambacorta et al., 2014). These differences in financial structure reflect some countries' characteristics such as the sectoral composition of output, with some productive sectors (e.g., agriculture and construction) that are more amenable to bank debt finance, as well as firm size, with small firms typically depending on bank finance because of the fixed costs involved in tapping capital markets, in addition to those associated with the corresponding governance mechanisms (Gambacorta et al., 2014). This is particularly true for the EU context where the prevalence of small and medium-sized enterprises (SMEs), which are the backbone of Europe's economy representing 99% of all businesses in the EU, makes the process of indirect finance (called financial intermediation) the primary route for moving funds from lenders to borrowers in most EU member states (ECB, 2022).

The literature does not favour one particular financial structure over the other. Indeed, the majority of studies show that it is the overall degree of financial development that matters for the real economy, and that both financial intermediaries and markets are important for economic growth (Beck & Levine, 2004; Boyd & Smith, 1998; Chakraborty & Ray, 2006; Demirgüç-Kunt & Levine, 2001, Levine, 2002). However, it is also worth noting that recent evidence suggests that their relative importance may vary depending on a country's level of economic and financial development (see, for example, Demirgüç-Kunt et al., 2011). This aligns with the goals of the Capital Market Union (CMU), as we will discuss in Chapter 3.

In the following paragraphs, the functioning and the main characteristics of the two finance channels will be outlined, with a particular focus on the EU context.

2.2 The Direct Financing Channel: Financial Markets

As outlined in the previous paragraph, in the *direct financing* channel deficit units borrow funds directly from lenders in financial markets by selling them financial instruments. Financial markets include any place or system that provides buyers and sellers with the means to trade financial instruments, including bonds, equities, the various international currencies and derivatives. Hence, they facilitate the interaction between economic actors who need capital (those who issue new securities or sell existing ones) with actors who have capital to invest (those who buy securities).

Based on the type of financial instruments traded, it is possible to identify four main categories of financial markets on which market participants can operate in order to meet their financial needs. These are the money, bonds, equity and derivatives markets. Before analysing the main characteristics of the abovementioned markets, let us introduce some general categorizations of financial markets in order to describe their essential features.

Debt and Equity Markets

Firstly, market participants can obtain funds in the financial market in two ways. The most common method is to issue a *debt instrument*, such as bonds, which is a contractual agreement by the borrower to pay the holder of the instrument fixed amounts at regular intervals (interest and principal payments) until a specified date (the maturity date), when a final payment is made. The second method of raising funds is by issuing *equities*, such as common stocks, which represent claims to share in the net income and the assets of a business. Equity instruments represent ownership in a company and make periodic payments (dividends) to their holders.

Primary and Secondary Markets

Secondly, the trading of financial instruments can occur in primary and secondary markets. A *primary market* is a financial market in which new issues of a security, such as bonds or stocks, are sold to initial buyers (investors) by a corporation or government borrowing funds. Initial

Public Offerings (IPOs) are an example of primary market.[2] They refer to the process by which a private company becomes a publicly listed company through the sale of a certain percentage of its stocks to the public, allowing a company to raise capital from public investors. On the other hand, a ***secondary market*** is a financial market in which securities that have been previously issued can be resold. The New York Stock Exchange (NYSE) and the London Stock Exchange are the best-known examples of secondary markets, where investors can buy and sell securities previously offered in the primary market. It is worth noting that the corporation (or government) that issues securities acquires new funds only when its securities are sold in the primary market. Indeed, when an investor buys a security in the secondary market, the actor who has sold the security receives money in exchange for the security, while the corporation that issued the security acquires no new funds.

Exchanges and Over-the-Counter Markets

Thirdly, there are two basic ways to organize financial markets, namely exchanges and over-the-counter (OTC) markets. *Exchanges* are organized marketplaces, typically a physical location, where buyers and sellers meet on a regular basis to trade securities using an open-outcry auction model. They are managed by an organization that sets the institutional rules that govern trading and information flows about the trading, thereby ensuring transparency of both pricing and transactions. Specifically, these rules concern the functioning of the market itself, the instruments and actors admitted to trading and the trading methods themselves. According to these regulations, every trade on an exchange is guaranteed and settled, usually through a clearinghouse. The clearinghouse is a market infrastructure which acts as counterparty in all transactions, thus eliminating the counterparty risk in exchange traded operations. Hence, its main role is to guarantee the successful completion of the transaction, i.e. the buyer receives the financial instrument he intends to acquire and the seller receiving the right amount paid for the financial instrument he is selling. Some of the best-known exchanges include the New York Stock Exchange, that is the world's largest stock exchange, the London Stock

[2] In this market, it is worth mentioning the important role played by investment banks in assisting in the initial sale of securities (see Chapter 4).

Exchange and the Italian Stock Exchange[3] (based in Milan) which are two of Europe's primary stock exchanges.

Unlike exchanges, *OTC* markets are not organized in the sense of having a physical location where trading takes place. Instead, trading occurs over sophisticated telecommunications networks: they are less formal, although often well-organized, networks of trading relationships centred around one or more dealers that act as market makers by quoting prices at which they will sell or buy securities to other dealers and to their customers. Hence, they are decentralized markets where geographically dispersed dealers are linked by telephones and computers. OTC markets are less regulated than exchanges, and this offers greater flexibility in making agreements that are specific to participants needs and goals. Indeed, trading arrangements are not standardised, and "atypical" contracts can be concluded. However, the unregulated nature of OTC markets makes them less transparent than exchanges, as well as characterized by a higher risk of a counterparty defaulting on any given agreement. The most popular OTC market is the Foreign exchange market (Forex) where currencies are bought and sold via a network of banks, instead of on exchanges.

It is worth mentioning that, in the past, the distinction between exchanges and OTC markets concerned the fact that exchanges were physical places where buyers and sellers meet on a regular basis to trade securities, as opposed to OTC markets where transactions occur electronically and in a decentralized way, i.e. directly between two parties. However, recent technological developments have blurred the traditional distinction, eliminating the need for exchanges to be physical places, and making electronic trading the most frequently used model. For instance, The NYSE currently advertises itself as a hybrid market that combines aspects of electronic trading and traditional auction-market trading, while an example of purely digital exchanges is the NASDAQ (National Association of Securities Dealers Automated Quotation System), the first electronic stock exchange founded in 1971.

Money and Capital Markets

[3] Since April 2021 Borsa Italiana stock exchange is part of Euronext, the pan-European stock exchange and market infrastructure.

Finally, another way of distinguishing between financial markets is on the basis of the maturity of the securities traded in each market. In particular, it is possible to differentiate between the *money market*, where short-term debt instruments (generally those with original maturity of less than one year) are traded, and the *capital market* where longer-term debt (generally with original maturity of one year or greater) and equity instruments are traded.

The description of the abovementioned categorizations of financial markets illustrates the general features of financial markets. The remainder of this paragraph provides a discussion of the money market, bonds and equity markets (as part of capital market), and derivatives market.

2.2.1 The Money Market

The money market involves transactions of securities that are short term, usually with a maturity of up to one year. The need for a money market arises because receipts of economic units do not coincide with their expenditures, i.e. revenues and expenses occurring at different times, generating liquidity problems. The money market provides an efficient, low-cost way of solving this problem, allowing lenders and borrowers to satisfy their short-term financial needs or, in other words, manage their cash and liquidity positioning (Hartmann et al., 2001).

Money markets are particularly essential to the liquidity management of banks, allowing liquidity to be readily transferred from banks with a surplus to banks with a deficit (Allen et al., 2009; De Fiore et al., 2018). Banks borrow and lend money in the money market also to satisfy regulations such as reserve requirements: if a bank cannot meet liquidity requirements, it will need to borrow money to cover the shortfall, while banks that have excess liquid assets above and beyond the liquidity requirements will lend money in order to earn interest on the assets. In addition to banks, other financial intermediaries such as insurance companies, investment and money market funds, as well as non-financial corporations, rely on money markets for their short-term funding needs (Corradin et al., 2020). Money markets are also key for the implementation and transmission of monetary policy. In particular, central banks (e.g., the European Central Bank [ECB], the Federal Reserve [FED], Bank of Japan [BoJ], etc.) are key players which operate on the money market for controlling the economy: central banks hold vast quantities of

Treasury securities that they sell if they believe the money supply should be reduced, or purchase Treasury securities if they believe the money supply should be expanded.

A variety of money market instruments are available to meet the diverse financial needs of market participants. They generally have a low default risk, being characterized by a high degree of safety of principal (most market participants are creditworthy actors) (Cook & LaRoche, 2010), while maturities range from one day to one year. According to the 2020 Euro money market study[4] by the ECB (2020a), which presents a comprehensive analysis of the functioning of euro money markets (see Box 2.1), the *secured segment* (i.e. transactions where the party providing liquidity receives collateral—like bonds—in return) is the largest segment of the euro money market, representing slightly more than two-thirds of total market turnover. This segment consists of daily *repurchase agreement* (also called Repo) transactions that are conducted by banks with financial corporations, general government and non-financial corporations. Repos are contracts in which a seller exchanges one or more securities[5] for money and, at the same time, commits to repurchase them at a future date by the same buyer at a predetermined price, which typically includes interest at an agreed-upon rate. In general, they have very short-term maturities: the majority of the transactions have a duration of one day, although significant volumes are traded also up to maturities of one/two months. They are also usually low-risk investments and therefore have low interest rates.

Box 2.1: The Euro Money Market

In this paragraph, we focused on the most popular money market instruments used by economic actors to manage their cash and liquidity positioning. However, it is worth noting that the euro money market also involves the following segments:

[4] The study describes developments in the Euro money market between January 2019 and December 2020. It relies predominantly on granular data collected through the Eurosystem's money market statistical reporting (MMSR) dataset, which contains details on volume, pricing, maturity and counterparty for each transaction of less than one year executed by the 48 largest euro area banks.

[5] Most Repos involve government bonds.

1. The *foreign exchange swaps* (FX swaps), which is the second-largest segment representing 22% of the total flows and 43% of the total stock of the euro money markets. It involves the borrowing and lending of foreign currencies, collateralized using domestic currency. FX swaps are widely used by banks and other financial institutions in order to hedge foreign currency exposures, arbitrage the basis and/or close liquidity gaps in foreign currency holdings (i.e. manage deficits or surpluses of foreign currencies) on the balance sheet.
2. The overnight *index swaps* (OIS) which consists of daily euro overnight index swap transactions denominated in euro of any maturity (it is the maturity of the underlying asset that qualifies the OIS as a money market instrument, regardless of the final maturity of the OIS), that are conducted with financial corporations, general government as well as with non-financial corporations. OIS contracts are used by market participants to manage interest rate risk and can be used to infer market expectations of the path of interest rates.

The *unsecured segment* (i.e. transactions without collateral), on the other hand, represented 14% of euro money market turnover in 2020, predominantly reflecting banks' borrowing transactions with non-banks. This segment includes all unsecured transactions, such as the issuance of commercial papers and certificates of deposit by various market players (e.g., banks and their underlying clients, corporations and public authorities). A *negotiable certificate of deposit* is a bank-issued security that documents a time deposit placed with a depository institution. Specifically, the certificate states the amount of the deposit, the date on which it matures, the interest rate (that can be fixed or variable) and the method under which the interest is calculated. This means that whoever holds the instrument at its maturity receives the principal and interest. However, it is worth mentioning that this instrument can be bought and sold until maturity. Negotiable certificates of deposit typically have a maturity of one to six months. *Commercial papers*, on the other hand, are unsecured debt instruments that are used by both financial and non-financial corporates to cover short-term funding needs. Their maturity is typically well below one year (no more than 270 days). Because these securities are unsecured, only the largest and most creditworthy corporations issue commercial papers. Their advantage is that they have a lower interest rate than bank credit. According to recent ECB (2020b) estimates, the market for corporate commercial papers denominated in euro currently amounts to approximately EUR75 bn, even if the depth of the commercial paper market differs widely across jurisdictions. The majority of euro area commercial paper is issued by monetary financial institutions, even if non-financial corporations have also become more active in the commercial papers market over time to manage their short-term cash needs.

2.2.2 Capital Markets

Economic actors that issue capital market securities and investors who buy them have very different motivations than those who operate in the money markets. The primary issuers of capital market securities are governments and both financial and non-financial corporations: they use capital markets for their long-term financial needs. For example, governments can issue long-term bonds to finance capital projects, such as infrastructure construction; similarly, non-financial corporations can collect the necessary financial resources to finance their growth and investment opportunities. Accordingly, by contrast to money market instruments, capital markets instruments have a maturity date over 1 year, or they do not have a maturity date.

Based on the type of financial instrument negotiated, we can identify two specific segments of capital markets: the bond and equity markets.

2.2.2.1 Bonds Market

Bonds are securities that represent a debt owed by the issuer to the investor. A bond obligates the issuer to pay a specified amount at a given date, generally with periodic interest payments. Notably, the corporation issuing the bond makes a legal commitment to pay interest on the principal and to return the principal when the bond comes due or matures. There are some key elements of a bond: (1) the *face value* (also called *par value*), that is the amount that the issuer must pay at maturity; (2) the *maturity date*; and (3) the *coupon rate*, that is the rate of interest that the issuer must pay periodically (it is the cost of the lending). However, it is worth mentioning that some bonds do not pay interest during their life. They are called zero coupon bonds: their remuneration is determined entirely by the discount at issue, i.e. the difference between the nominal value (i.e. face value) and the price paid.

Bonds can be issued by central or local governments (government bonds), as well as by both financial and non-financial corporations (corporate bonds).

Government bonds are the main instruments that euro area governments use to finance their national debts. The proceeds from these bonds are used to finance public interest projects, such as schools, utilities and transportation systems. Their maturity is over 1 year: for example, Italian BTPs are issued with maturities 5, 10, 15 and 30 years, while German Federal bonds (BUND) with maturities of 7–30 years. It is worth mentioning that governments also issue short-term bonds, i.e. with a maturity of one year or less. An example of these instruments is the Italian Treasury Bills (BOTs). They are short-term securities with maturities up to one year (3/6/12 months or any other maturity within one year). Short-term government bonds are money market instruments. Government bonds are generally considered to be free of default risk since there is a low chance that a government will default on its debt. Default risk, also called default probability, is the probability that a borrower fails to make full and timely payments of principal and interest, according to the terms of the debt security involved.

Corporate bonds, on the other hand, are debt instruments issued by both financial and non-financial corporations to collect the necessary financial resources to finance their growth. For example, non-financial corporations use the proceeds from bond sales for a wide variety of purposes, including buying new equipment, investing in research and development, refinancing debt, and financing mergers and acquisitions. Maturities can be medium term (over 1–10 years) or long term (more than 10 years). The degree of risk of corporate bonds varies widely among different bond issuers because the risk of default depends on the company's financial health, which can be affected by a number of economic and financial variables (liquidity, solvency, profitability and operating efficiency). Accordingly, the interest rate on corporate bonds varies with the level of risk of the company.

There are several categories of corporate bonds. Below, we describe the most popular:

1. *Secured* and *Unsecured bonds*: secured bonds are those with a collateral attached. For example, a building may be the collateral for bonds issued for its construction. In the event that the firm fails

to make payments as promised, bondholders have the right to liquidate the property in order to be paid. Because these bonds have a collateral, they are less risky than comparable unsecured bonds. Conversely, unsecured bonds are those that are backed only by the general creditworthiness of the issuer. Hence, no specific collateral is pledged to repay the debt. As a result, they will have a higher interest rate than otherwise comparable secured bonds.

2. *Subordinated bonds* are unsecured bonds that have a lower priority claim, meaning that in the event of a default, subordinated bondholders are paid only after non-subordinated bondholders have been paid in full. Therefore, they are considered riskier.
3. *Convertible bonds* and *bonds cum warrant*: A convertible bond is a type of debt security that provides an investor with a right to exchange the bond for a predetermined number of shares in the issuing company at certain times of a bond's lifetime. This enables bondholders to become a shareholder of the company and acquire all rights that stocks provide. The number of stocks depends on the face value of the bond, as well as the price of stocks. The main elements of such a type of bond are: (1) the *conversion method*: direct (if the convertible shares are issued by the same issuer as the bonds), or indirect (if the convertible shares are issued by a company other than the issuer of the bonds); (2) the *conversion price* (or conversion ratio), which expresses the number of shares obtainable for each bond; (3) the *conversion period*, which represents the period(s) from which conversion may be requested.

   Similarly, bonds cum warrant give investors the right to purchase a certain amount of shares at a fixed price. However, cum warrant bonds differ from convertible bonds for the following reasons: firstly, this right is embedded in an instrument, the *warrant*, which can be detached from the bond and circulate independently in the market; secondly, the bond continues to exist following the exercise of the warrant, whereas the convertible bond ceases to exist upon conversion; thirdly, an additional cash outlay is required to purchase the conversion shares, with respect to the amount already invested in the bond, equal to the predetermined exercise price multiplied by the number of conversion shares.

Long-term debt securities issued by public authorities represent the most important segment of the Eurozone bond market, closely followed by the financial intermediaries segment—especially monetary financial institutions, while non-financial corporations account for a small share of the market because of their leverage on bank loans.[6] However, according to the ECB (2022) estimates, firms in the euro area have increasingly resorted to bond financing, especially following the global financial crisis (GFC) of 2008–2009, with the outstanding volume of bonds relative to bank borrowing by euro area firms that has risen to around 30%, up from roughly 15% in mid-2008 (Holm-Hadulla et al., 2022).

2.2.2.2 The Equity Market

Stocks represent an ownership interest—or equity—in a corporation. The equity market gives companies access to capital to grow their business, and investors a piece of ownership in a company with the potential to realize gains in their investment based on the company's future performance. In contrast to bonds, ownership of stock gives investors certain rights regarding the firm. Firstly, stocks give investors economic rights in terms of dividend payments. Dividends are distributions of earnings made by a corporation to its shareholders. They represent a return on investment in equity securities, corresponded in the form of cash or additional shares of stocks (stock dividends). This is because stockholders have a claim on all assets and income left over after all other claimants have been satisfied (the right of a *residual claimant*). In addition, certain types of stocks give investors administrative rights, that is the right to vote during shareholder meetings for directors and on other business issues, such as amendments to the corporate charter and whether new shares should be issued.

There are several categories of stocks, among which the most common used is between common and preferred stocks. *Common stocks* give both administrative and economic rights, meaning that common shareholders can vote and receive dividend payments. On the other hand, *preferred stocks* do not give voting rights. Therefore, when a company has to elect a board of directors or vote on any form of corporate policy, preferred shareholders have no voice in the future of the company. Nevertheless, preferred stockholders have several economic advantages compared to common stockholders: they generally receive a fixed dividend; they also hold a claim on assets that has priority over the claims of common shareholders but after that of creditors such as bondholders.

[6] See, for more details, the Securities issues statistics (SEC) by the ECB.

Fig. 2.2 Market capitalization of listed domestic companies (% of GDP) (*Source* Own processing of data from the World Bank)

It is worth noting that in the EU context, equity markets play a limited role as a source of new funds for corporations compared to other major advanced economies, such as the USA and Asia. As Fig. 2.2 shows, the size of EU equity markets, measured by the market capitalization of companies listed in each country as a percentage of their GDP, is lower compared to those of the USA and East and Pacific countries (Data for the EU is available until 2018). As discussed before, this is because in the EU context debt financing via bank loans is more important than financing via listed shares.

2.2.3 *Derivatives Market*

A financial market that has experienced quick growth all over the world in the last two decades is the derivatives market. According to the latest statistics collected by the Bank for International Settlements (BIS) (2022), the notional value of outstanding OTC derivatives rose to $632 trillion at end-June 2022, up from $598 trillion at end-2021, while the gross market value of outstanding OTC derivatives, summing positive and negative values, rose noticeably in the first half of 2022, to $18.3 trillion.

Derivatives are financial instruments whose performance is derived, at least in part, from the performance of an underlying asset, which can be a financial asset (including stocks, bonds), market indices, interest or exchange rates, as well as commodities (such as natural gas, gold, oil, etc.). There are four main categories of derivative contracts. These different types of contracts entail private agreements between two parties to exchange future cash flows according to a predetermined formula (*swaps*), to buy or to sell an asset at a certain future time at a fixed price (*forward* and *futures*),[7] or the right to buy or to sell an asset at any time up to a given expiration date (*options*).

The primary users of derivatives are financial intermediaries such as banks and insurance companies (Berends & King, 2015; Infante et al., 2020). However, the use of financial derivatives by non-financial corporations has also grown over the years (Bartram et al., 2009). These market participants use derivatives for purposes that are different from those related to the other financial markets investigated before. Notably, derivatives can be used as a source of revenue, being used for trading and broker-dealer purposes, but also as important risk management tools (hedging purposes) since they allow parties to identify, isolate and manage the market risk associated with financial instruments and commodities, such as change in the market prices of financial instruments and shifts in interest or exchange rates, thereby reducing the degree of financial risk to which they are exposed (De Haan et al., 2020; Géczy et al., 1997; Infante et al., 2020; Minton et al., 2009; Nguyen & Faff, 2010).

Nonetheless, it is worth mentioning that the use of derivatives has raised several concerns among policymakers regarding their effects on the performance and risk of firms, particularly due to the high degree of complexity of some derivative contracts and their frequent use in the OTC market. Notably, building on the idea that banks could use derivatives for non-hedging purposes, many authors have argued that derivatives usage could tempt bank managers to engage in excessive risk-taking (Franke & Krahnen, 2008; Instefjord, 2005) and lead to a destabilizing concentration of risks (Rajan, 2006; Stulz, 2004), which in turn may undermine the stability of the banking system (Dewally & Shao, 2013; Trapp & Weiß, 2016).

[7] The difference between futures and forwards is that *futures* are standardized contracts, meaning they are traded on the exchange market; by contrast, *forwards* contracts are over-the-counter products, which means they are traded over the counter and not on the exchange market. Accordingly, forwards contracts are unstandardized: they are customizable to suit the requirements of both parties involved.

2.3 The Indirect Financing Channel: Financial Intermediaries

In addition to financial markets, funds can also move from lenders to borrowers through a financial intermediary that helps transfer funds from one party to the other. This process is called *financial intermediation*. In order to understand the important role played by financial intermediaries in the financial system, it is useful to recall the traditional *theories of financial intermediation*, which explain why intermediaries and indirect finance are so important for the well-functioning of the financial system. These theories build on the notion that intermediaries serve to overcome market imperfections, which prevent savers and investors from trading directly with each other in an optimal way, thereby facilitating the channelling of funds from surplus units to deficit units (Allen & Santomero, 1997; Merton, 1995; Pyle, 1971; Scholtens & van Wensveen, 2003). Notably, these market imperfections are transaction costs, informational asymmetries and high uncertainty, which are discussed briefly below.

Transaction costs refer to the time and money required to carry out financial transactions. These include, for example, search and information costs (i.e. costs associated with the searching activity for a counterparty and those associated with acquiring all relevant information), as well as the costs of writing and enforcing contracts. When high, these costs may make the exchange expensive for one or both parties involved in a transaction, thereby hindering the channelling of funds from lenders to borrowers. According to the financial intermediary theories, financial intermediaries can substantially reduce such costs, and consequently, they can facilitate the transferring of funds from lenders to borrowers. This is because financial intermediaries have developed expertise in lowering transaction costs, being specialized in investments, and because they can take advantage of economies of scale, i.e. when unit costs decrease as production volume increases, or in other words, when a bank reduces the average cost of production when increasing the level of output.

The problem of *asymmetric information* is acknowledged as an additional reason for the existence of financial intermediaries. Asymmetric information appears when one party in a transaction is more (or better) informed than the other. For example, a borrower who takes out a loan usually has better information about the potential returns and risks associated with the investment projects for which the funds are earmarked than the lender does. Accordingly, this situation points out the advantageous position of the deficit unit compared to the surplus unit. Information asymmetries raise problems before the transaction occurs (i.e. ex-ante information asymmetry) for the initial selection of borrowers, as well as after the transaction takes place (ex-post information asymmetry) for monitoring the behaviour of borrowers in the use of the financial resources received, which are an important impediment to well-functioning financial markets.

Ex-ante information asymmetry arises because borrowers generally know more about their investment projects than lenders while, at the same time, it is difficult and costly for lenders to evaluate potential borrowers since they may not have the time, capacity or means to collect and process information on a wide array of potential borrowers. This situation may lead to ***adverse selection*** problems, which occur when the potential borrowers who are the most likely to produce an undesirable (***adverse***) outcome—bad credit risks—are the ones most likely to seek the transaction and be selected. Because adverse selection makes it more likely that loans might be made to bad credit risks, lenders may decide not to make any loans even though there are good credit risks in the marketplace. Accordingly, high information costs may prevent funds from flowing from lenders to borrowers. On the contrary, financial intermediaries can overcome problems associated with adverse selection since they have a better way of selecting the good credits, thereby improving resource allocation. This is because financial intermediaries can reduce the costs of gathering and processing information thanks to the economies of scale and cost savings also from the use of digital technology.

Additionally, there can be ex-post information asymmetry problems when borrowers have incentives to engage in undesirable activities once the transaction has taken place. This is called a ***moral hazard*** problem, that is once a loan has been granted, there can be the risk that the borrower will engage in undesirable activities (from the perspective of the lender) that increase the probability of default. Because moral hazard lowers the probability that the loan will be repaid, lenders may decide that they would rather not make a loan. Again, financial intermediaries can alleviate these problems as they mitigate the information acquisition and costs of monitoring borrowers. For example, when granting a loan to a firm, banks can create arrangements that force managers to behave in a way that ensures that the loan will be repaid.

In addition to information and transaction costs, some scholars have identified other reasons to explain the existence of financial intermediaries and their important role in the modern economy. In this respect, it useful to mention Scholtens and Van Wensveen (2000) who claimed that the origins of intermediaries (banking and insurance) lie in their *risk transforming* and *risk managing* functions, which help reduce the exposure of investors to risk (e.g., counterparty risk, market risk, etc.), that is, uncertainty about the returns investors will earn on assets. In particular, according to the authors financial intermediaries do this through the process known as *asset transformation*, that is the process of turning risky assets into safer assets for investors. In other words, financial intermediaries create and sell assets with risk characteristics that people are comfortable with, and then use the funds acquired by selling these assets to purchase other assets that may have far more risk. Further, financial intermediaries can perform their risk management function by helping individuals to diversify their investments, i.e. investing in a collection (portfolio) of assets whose returns do not always move together, with the result that the overall risk is lower than for individual assets, thereby lowering the amount of risk to which investors are exposed. This process is called *diversification*.

Overall, the considerations above suggest that financial intermediaries play an important role in the economy, because they lower transaction costs, they overcome information problems and promote risk sharing, thereby helping financial markets channel funds from lender-savers to economic actors with productive investment opportunities.

2.3.1 Financial Intermediaries

In the previous paragraph, we have explained why financial intermediaries play such an important role in the economy. Here, we provide an overview on the role and the business of the principal financial intermediaries. They fall into different categories as defined by the ECB.

A first categorization concerns the difference between *monetary financial institutions* (MFIs) and *other financial intermediaries* (OFIs). MFIs include all credit institutions and non-credit institutions (mainly money market funds) whose business is to receive deposits from entities other than MFIs and to grant credit and/or invest in securities.[8] *Commercial banks* are among the most important monetary financial institutions. Their main business is *lending*: they raise funds primarily by issuing deposits, and then use these funds to make commercial, consumer and mortgage loans. The difference (or spread) between the lending and borrowing interest rates is the most important determinant of a commercial bank's profitability. On the other hand, *other financial intermediaries* (OFIs) include all corporations other than insurance corporations and pension funds that are engaged mainly in financial intermediation by incurring liabilities in forms other than currency, deposits and/or close substitutes for deposits from institutional entities other than MFIs, in particular those engaged primarily in long-term financing, such as corporations engaged in financial leasing, financial holding corporations and venture capital corporations, among others.

Among the different financial intermediaries, insurance corporations and investments funds emerge as important players in the EU context.

Insurance corporations offer direct insurance or reinsurance services, providing financial protection from possible hazards in the future. Fundamentally, they are in the business of assuming risk on behalf of their customers in exchange for a fee (called a *premium*), meaning that the insurance corporation undertakes to compensate the policyholder for losses caused by a pre-defined event. Insurance corporations make a profit by charging premiums that are sufficient to pay the expected claims to the company plus a profit. Typically, insurance corporations may cover specific kinds of events:

- *Life insurance* policies which protect against risks involving human life: the event is usually the death or a deterioration in the health of the insured person. Here, policyholders make regular or one-off payments to the insurer in return for which the insurer guarantees to give policyholders an agreed sum, or an annuity, at a given date or earlier.

[8] MFIs also include the Eurosystem, i.e. the ECB and the National Central Banks of those countries that have adopted the euro.

- *Non-life insurance* policies protect against risks of financial losses. They cover expenses the policyholder incurs from damages to health or property (policies typically offered are medical expenses, or house, motor vehicle and fire insurance), and financial losses like a loss of income.

Investment funds (IFs),[9] on the other hand, are collective investment undertakings that invest in financial and non-financial assets to the extent that the objective is to invest capital raised from the public. Investment funds can be distinguished by investment policy (i.e. equity funds, bond funds, mixed funds, real estate funds, hedge funds, other funds) and by type of fund, namely open-end vs closed-end funds. *Open-end funds* are undertakings whose units or shares are, at the request of the holders, repurchased or redeemed directly or indirectly out of the undertaking's assets; on the other hand, *closed-end funds* are undertakings which have a fixed number of issued shares and whose shareholders have to buy or sell existing shares when entering or leaving the fund.

Although banks continue to play a primary role in the Eurozone financial system, the latest ECB's (2022) biennial report on financial integration and structure in the euro area shows a softening of bank dominance and the increasing weight of non-bank financial intermediaries (e.g., money market funds, investment funds and insurance corporations, among others),[10] which continue to gain importance for euro area real economy financing. To illustrate, the ratio between credit granted by non-banks to non-financial corporations to credit provided by both banks and non-bank financial intermediaries has almost doubled from around 15% to close to 30% since the global financial crisis.

[9] Money market funds and pension funds are not covered in this category.

[10] They held a combined €31.7 trillion in financial assets by end-2021 and accounted for 37.3% of the total financial sector.

References

Allen, F., Carletti, E., & Gale, D. (2009). Interbank market liquidity and central bank intervention. *Journal of Monetary Economics, 56*(5), 639–652.

Allen, F., & Santomero, A. M. (1997). The theory of financial intermediation. *Journal of Banking & Finance, 21*(11–12), 1461–1485.

Bank of International Settlements. (2022). *OTC derivatives statistics at end-June 2022*. https://www.bis.org/publ/otc_hy2211.pdf

Bartram, S. M., Brown, G. W., & Fehle, F. R. (2009). International evidence on financial derivatives usage. *Financial Management, 38*(1), 185–206.

Beck, T., & Levine, R. (2002). Industry growth and capital allocation: Does having a market-or bank-based system matter? *Journal of Financial Economics, 64*(2), 147–180.

Beck, T., & Levine, R. (2004). Stock markets, banks, and growth: Panel evidence. *Journal of Banking & Finance, 28*(3), 423–442.

Beck, T., Levine, R., & Loayza, N. (2000). Finance and the sources of growth. *Journal of Financial Economics, 58*(1–2), 261–300.

Berends, K., & King, T. B. (2015). Derivatives and collateral at US life insurers. *Federal Reserve Bank of Chicago Economic Perspectives, 39*(1), 21–37.

Boot, A. W., & Thakor, A. V. (1997). Financial system architecture. *The Review of Financial Studies, 10*(3), 693–733.

Boyd, J. H., & Smith, B. D. (1998). The evolution of debt and equity markets in economic development. *Economic Theory, 12*, 519–560.

Chakraborty, S., & Ray, T. (2006). Bank-based versus market-based financial systems: A growth-theoretic analysis. *Journal of Monetary Economics, 53*(2), 329–350.

Cook, T. Q., & LaRoche, R. K. (2010). The money markets. In *Introduction to financial markets and institutions* (p. 239). Routledge.

Corradin, S., Eisenschmidt, J., Hoerova, M., Linzert, T., Schepens, G., & Sigaux, J. D. (2020). Money markets, central bank balance sheet and regulation. *SSRN Journal*. https://doi.org/10.2139/ssrn.3716200

De Fiore, F., Hoerova, M., & Uhlig, H. (2018). *Money markets, collateral and monetary policy* (No. w25319). National Bureau of Economic Research.

De Haan, J., Schoenmaker, D., & Wierts, P. (2020). *Financial markets and institutions: A European perspective* (4th ed.). Cambridge University Press.

Demirgüç-Kunt, A., Feyen, E., & Levine, R. (2011). *Optimal financial structures and development: The evolving importance of banks and markets*. World Bank, mimeo.

Demirgüç-Kunt, A., & Levine, R. (Eds.). (2001). *Financial structure and economic growth: A cross-country comparison of banks, markets, and development*. MIT Press.

Demirgüç-Kunt, A., & Maksimovic, V. (2002). Funding growth in bank-based and market-based financial systems: Evidence from firm-level data. *Journal of Financial Economics, 65*(3), 337–363.

Dewally, M., & Shao, Y. (2013). Financial derivatives, opacity, and crash risk: Evidence from large US banks. *Journal of Financial Stability, 9*(4), 565–577.

European Central Bank. (2020a). *Euro money market study*. https://www.ecb.europa.eu/pub/euromoneymarket/html/ecb.euromoneymarket202104_study.en.html

European Central Bank. (2020b). *The ECB's commercial paper purchases: A targeted response to the economic disturbances caused by COVID-19*. https://www.ecb.europa.eu/press/blog/date/2020/html/ecb.blog200403~54ecc5988b.en.html

European Central Bank. (2022, April). *Financial integration and structure in the Euro area*. https://www.ecb.europa.eu/pub/pdf/fie/ecb.fie202204~4c4f5f572f.en.pdf

Franke, G., & Krahnen, J. P. (2008). *The future of securitization* (CFS working paper No. 2008/31).

Gambacorta, L., Yang, J., & Tsatsaronis, K. (2014, March). Financial structure and growth. *BIS Quarterly Review*.

Géczy, C., Minton, B. A., & Schrand, C. (1997). Why firms use currency derivatives. *The Journal of Finance, 52*(4), 1323–1354.

Hartmann, P., Manna, M., & Manzanares, A. (2001). The microstructure of the euro money market. *Journal of International Money and Finance, 20*(6), 895–948.

Holm-Hadulla, F., Musso, A., Nicoletti, G., & Tujula, M. (2022). Firm debt financing structures and the transmission of shocks in the euro area. *Economic Bulletin Articles, 4*. https://www.ecb.europa.eu/pub/economic-bulletin/articles/2022/html/ecb.ebart202204_02~b35a8321b7.en.html

Infante, L., Piermattei, S., Santioni, R., & Sorvillo, B. (2020). Diversifying away risks through derivatives: An analysis of the Italian banking system. *Economia Politica, 37*, 621–657.

Instefjord, N. (2005). Risk and hedging: Do credit derivatives increase bank risk? *Journal of Banking & Finance, 29*(2), 333–345.

Levine, R. (2002). Bank-based or market-based financial systems: Which is better? *Journal of Financial Intermediation, 11*(4), 398–428.

Levine, R., Loayza, N., & Beck, T. (2000). Financial intermediation and growth: Causality and causes. *Journal of Monetary Economics, 46*(1), 31–77.

Levine, R., & Zervos, S. (1998). Stock markets, banks, and economic growth. *American Economic Review, 88*, 537–558.

Merton, R. C. (1995). A functional perspective of financial intermediation. *Financial Management, 24*, 23–41.

Merton, R. C., & Bodie, Z. (1995). A conceptual framework for analyzing the financial system. In *The global financial system: A functional perspective* (pp. 3–31). Harvard Business School Press.

Minton, B. A., Stulz, R., & Williamson, R. (2009). How much do banks use credit derivatives to hedge loans? *Journal of Financial Services Research, 35*, 1–31.

Mishkin F. S. & Eakins S. (2012). *Financial markets and institutions* (Global Edition, 7th ed.). Pearson.

Mishra, S., & Narayan, P. K. (2015). A nonparametric model of financial system and economic growth. *International Review of Economics & Finance, 39*, 175–191.

Nguyen, H., & Faff, R. (2010). Are firms hedging or speculating? The relationship between financial derivatives and firm risk. *Applied Financial Economics, 20*(10), 827–843.

Pyle, D. H. (1971). On the theory of financial intermediation. *The Journal of Finance, 26*(3), 737–747.

Rajan, R. G. (2006). Has finance made the world riskier? *European Financial Management, 12*(4), 499–533.

Rajan, R. G., & Zingales, L. (1996). *Financial dependence and growth*. https://www.nber.org/papers/w5758

Scholtens, B., & Van Wensveen, D. (2000). A critique on the theory of financial intermediation. *Journal of Banking & Finance, 24*(8), 1243–1251.

Scholtens, B., & Van Wensveen, D. (2003). *The theory of financial intermediation: An essay on what it does (not) explain* (No. 2003/1). SUERF Studies.

Stulz, R. M. (2004). Should we fear derivatives? *Journal of Economic Perspectives, 18*(3), 173–192.

Thakor, A. V., & Boot, A. (Eds.). (2008). *Handbook of financial intermediation and banking*. Elsevier.

Trapp, R., & Weiß, G. N. (2016). Derivatives usage, securitization, and the crash sensitivity of bank stocks. *Journal of Banking & Finance, 71*, 183–205.

Xu, Y., Zhao, X., Zhou, X., & Yuan, L. (2021). Financial structure and capital allocation efficiency: An empirical analysis of China from 2005–2018. *Applied Economics, 53*(29), 3347–3358.

CHAPTER 3

The European Banking System: Main Features Along the Evolutionary Path

Abstract This chapter outlines the evolution of the European integration process, giving an overview of the most important regulatory steps related to the banking and financial system culminated in the creation of the Banking Union. A focus is also given to the international financial crisis; its origins and impacts from the perspective of the European banking system and to the role played by supervision aimed at ensuring the financial stability.

Keywords European Union · International financial crisis · European banking system · Banking Union · Financial integration · Banking supervision · Capital adequacy

3.1 The European Union at a Glance

The European Union (EU) origins date back to the 1950s when some European countries (Germany, France, Italy, the Netherlands, Belgium and Luxembourg) formed the European Coal and Steel Community. This Community aimed at bringing together the six countries' coal and steel industries under common management. In this way, no country could be able on its own to manufacture war weapons to use against the others, as instead had happened in previous years (Glockner & Rittberger, 2012;

P. Ferretti and P. Martino, *Banking and Financial Markets*,
https://doi.org/10.1007/978-3-031-32562-5_3

Phinnemore, 2013). In other words, it represented the first step towards establishing real peace within Europe.

In 1958, it became the European Economic Community and in 1993 the European Union. Over the years, 22 countries have joined the founding members and in 2020 the United Kingdom left the EU. Currently, the EU members are 27[1] and its powers have increased significantly (EC, 2022a).

The integration process of the European Union has been characterized by important events. In the following, we summarize some of the most relevant steps of its evolution (Risenfeld, 1992).

In 1958, as mentioned above, under the Treaty of Rome the European Economic Community (EEC) was created in order to drive Europe towards a strong cooperation for integration and economic growth through trade. The goal was to create a common market based on the free movement of goods, people, services and capital (Healey, 1995; Phinnemore & Warleigh-Lack, 2009).

Some decades later, in 1985, The European Commission[2] published a white paper on the Completion of the Internal Market (EC, 1985). The objective was to review the EEC created under the Treaty of Rome in order to relaunch European integration and complete the internal market that is a zone without borders where there is free movement of goods, people, services and capital, by 1 January 1993. Moreover, rules about the functioning of the European institutions were revised and the powers of the EEC were extended. Thereby, the way was opened to further political integration and to the economic and monetary union, which would have been sanctioned by the Maastricht Treaty, officially the Treaty on the European Union, which laid the foundations for the current EU (EP, 1992).

Signed in Maastricht in February 1992, the Treaty on the European Union entered into force on 1st November 1993. Under it the

[1] The EU member States are the following: Austria, Belgium, Bulgaria, Croatia, Cyprus, Czechia, Denmark, Estonia, Finland, France, Germany, Greece, Hungary, Ireland, Italy, Latvia, Lithuania, Luxembourg, Malta, Netherlands, Poland, Portugal, Romania, Slovakia, Slovenia, Spain and Sweden.

[2] The European Commission is the EU's main institution and the only one able to submit proposals for new laws. In other words, it can initiate the formal legislative process by presenting a proposal to the European Parliament and the Council of the EU. Additionally, it manages EU policies and budget, and it ensures the adoption of EU regulations by all the countries.

EEC became the Economic Community and the powers of the European Parliament were enhanced. It also covered relevant topics. First of all, the enhancement of cooperation among European countries. To this end, it introduced European citizenship allowing people to choose the state of residence and move freely within the EU; it set up a common foreign and security policy aimed at safeguarding the common values, the key-interests and the independence of the EU; it developed a strong cooperation on justice and home affairs in order to ensure the security of European citizens (EP, 1992, 1999).

Furthermore, the Maastricht Treaty created the conditions for a single European currency: the euro. It established the European Central Bank (ECB) and the European Central Bank System (ECBS, formed by the ECB and the National Central Banks of all the EU countries regardless of the adoption of the euro). In short, it was the culmination of several decades of debate on the enhancement of economic cooperation within Europe. The previous idea of a single currency was based on the realization of a transitional three-step process and the Treaty of Maastricht formally defined such steps (together with the related time period): (*i*) free movement of capital (01.07.1990–31.12.1993); (*ii*) stronger cooperation among national central banks and greater alignment of economic policies of the member States (01.01.1994–31.12.1998); and (*iii*) a gradual introduction of the euro and the implementation of a single monetary policy by the ECB (01.01.1999-to date). The Maastricht Treaty has also regulated the practical aspects of the functioning of the euro, by setting the standards the countries should comply with for euro area membership (EP, 1992). Such standards aim to maintain the price stability within the euro area following the entry of new countries.

The start of the euro area is the 1 January 1999 when the National Central Banks of 11 EU countries transferred their monetary powers to the ECB, which became the monetary authority for the euro area (formed by all the members adopting the euro) (EC, 2022b). Since then, the ECB has been the only authority responsible for the common monetary policy, as EU countries decided to voluntarily share a currency and pool their monetary sovereignty. Price stability is the main objective of the ECB, which is established by maintaining low inflation rates in the medium term. The ECB together with the National Central Banks (responsible for monetary policy implementation) of the member States adopting the euro forms the Eurosystem (ECB, 2007; Moutot et al., 2008). For the first three years, the euro was a virtual currency, as it was only used by banks and financial markets, while on 1 January 2002 euro notes and coins were introduced (ECB, 2007).

More recently, in 2014, the European Banking Union started. It was the result of the crisis of previous years, particularly the international financial crisis of 2008 (Sect. 3.2) and the related sovereign debt crisis. Such events made it clear that the close connections between the public, finance and banking sectors could cross boundaries and negatively affect other countries within the EU. The Banking Union is a very important step of the European integration process. It provides the application of EU banking rules in order to ensure that the European banking market is more transparent, unified and safer (Gortsos, 2017; Véron, 2013).

Additionally, the promotion of the development of an integrated capital market in Europe (Capital Markets Union, see Sect. 3.3) was adopted in 2015 (ECA, 2020; Quaglia et al., 2016; Véron & Wolff, 2016).

3.2 The International Financial Crisis in the Perspective of the European Banking System

During the years before the start of the financial crisis (2008), the economies (particularly the USA) were characterized by certain features: (*i*) low levels of interest rates, due to expansionary monetary policies and significant capital inflows particularly from Asia, low inflation expectations and limited perception of default risk; (*ii*) sustained economic growth (especially for China and India); (*iii*) default rates for bonds and loans at historic lows with a downward trend; and (*iv*) credit spreads (that is the differential between loan rates and risk-free investment rates) at low levels (Cukierman, 2013; Ramskogler, 2014).

In particular, the low interest rates fostered investments and economic growth, that in turn helped firms grow and made defaults less likely. Low default rates decreased the risk-premium requested by professional investors (i.e. banks, funds and insurance companies) and therefore firms and households could borrow at lower rates. The resulting favourable environment that had been created fuelled a high liquidity level of the financial markets on an international scale (Blundell-Wignall et al., 2009; Colloquium & Papademos, 2010; McKibbin & Stoeckel, 2010).

In such a context, as mentioned, low interest rates conducted to economic development and to a consequent fast growth of credit supply, especially for residential mortgages, which promoted the property price boom (DiMartino & Duca, 2007; Duca et al., 2010; Goodhart, 2008). Starting around 2001 the USA housing market registered a boom and banks granted subprime mortgages that are housing loans to borrowers with a very high risk. Such subprime mortgages were attractive for banks for the opportunity to charge a relatively high interest rate in front of very low default rates due to the price boom (Acharya & Richardson, 2009; Allen & Carletti, 2010).

Additionally, the expansion of credit, even if in the presence of bad quality borrowers, was possible thanks to the disposal of mortgage loans and/or the associated credit risk by banks to the market: this is the practice of securitization (for an overview of its basic scheme of functioning see Box 3.1). Indeed, the great supply of mortgages was accompanied by an inappropriate credit risk assessment, as the related risks were transferred to the market thanks to securitization (Deku et al., 2017; Solomon, 2012).

Box 3.1: Securitization: A Short Overview on the Rationale and Functioning

Securitization represents a form of financial innovation. It started in the 70s in the US market where many loans (e.g., mortgages) were securitized as Asset-Backed-Securities (ABS). The securities were sold on the financial market in competition with other financial instruments

During the 90s and at the beginning of the 2000s the development of securitization increased and several types of ABS were issued, especially in the mortgage market. The final goal of securitization was to allow banks to transfer risks to the market and investors to have access to investments difficult to access on their own.

Indeed, securitization is based on the conversion of loans into marketable assets through the issue of securities transferable to the market. Selling the loans allows the originator to mobilize, before the expiration date, such assets which generate cash inflows, constant or predictable. It is well known indeed that banks finance their loans (e.g., mortgages) through their funding (e.g., deposits) and keep the loans (and the related risk) on their balance sheet until they expire. Such business generates interest income and fees for banks. Thanks to securitization on the contrary banks have the opportunity to resell the loans to the market. To this end, the bank-originator has to identify the loans to be securitized in order to sell them to a trust (the Special Purpose Vehicle—SPV), which buys a pool of loans of similar types with the aim of selling them to the market. Particularly, the SPV finances the purchase of loans by issuing securities (ABS) which are then placed on the market (generally with support from an investment bank).

It should be noted that the ABS principal and interest payment is closely related to the cash inflows of the loans disposed. Hence, the investors take the risk that the underlying loans do not generate an adequate cash flowing able to meet the debt service. In order to mitigate such risk there are some forms of credit enhancement. Moreover, the rating agencies are generally involved in order to make an assessment of the quality of the securities issued (that is their risk). So, before the securities come to the market, the rating agencies place their opinion on the deals (see for further details Gorton & Metrick, 2013).

At the same time, it is worth noting that some macro-economic conditions actually already showed elements of weakness

Indeed, the market was characterized by a high leverage; particularly in the US market, the liquidity excess increased the level of indebtedness of firms (due to the modest costs), households (interested in real estate investments and consumptions), private equity funds (engaged in acquisitions with a high debt-to-equity-ratio) and hedge funds (in the absence of regulation their business was characterized by high leverage for financing extraordinary operations) (Blundell-Wignall, 2007; Myers & Rajan, 1998).

Second, as above mentioned, the risks linked to loans disposed to the market were assessed inappropriately. Many international banks, in the face of high liquidity and modest margins linked to the low credit spreads, searched for profitability by intervening on the capital. It was not a mere increase of leverage, but an expansion of credit supply through the disposal of assets (e.g., loans) and/or credit risk, which allowed banks also to free up capital. Securitization is a good technique in itself, but it could create problems when it becomes practice and the quota of loans disposed reaches a very high percentage (the 80–90%) of the total loans (Taureck, 2006). There is also to consider that banks not only relied on securitization but they also invested in structured products resulting from securitization: the ABSs and Collateralized-Debt Obligations (CDOs) (Lucas et al., 2006). The CDOs issue is linked with a securitization thanks to which assets (e.g., loans) are disposed from an intermediary to a Special Purpose Vehicle (SPV). The latter, in face of the purchase of the pool raises funds issuing CDOs (commonly the issue is secured by the loans that have been disposed). The SPV generally issues different CDOs tranches, each one characterized by a specific risk and priority as regards reimbursement: a senior tranche is refunded (principal and interest) first, then the junior/mezzanine tranche and lastly the subordinated/equity tranche.

In line with the above-mentioned considerations, it is important to underline the insufficient attention given to the credit risk related to the securities resulting from securitization (ABSs and CDOs). The structured securities resulting from securitization were not appropriately assessed in terms of risk, specifically credit risk linked to the downgrading of the quality of the loans given as guarantee.

Around the mid-2000s changes of the macro-economic conditions took place: the increase of the prices of raw materials, the rise in inflation and interest rates and the decrease of real estate prices. In 2004, the Federal Reserve (the US Central Bank) started to raise interest rates with the aim of cooling down the economy and controlling inflation. The result was that the increase in mortgage rates led to the slowing down of the housing market and consequently to the collapse of housing prices (Bernanke, 2008). Particularly, the decrease of the housing value and the increase of the rates led to situations where the property value could be lower than the debt value and to growing mortgage payments. So, the basic assumption under which the subprime mortgages could be sold because the housing prices would have continued to increase failed. With the decrease of the housing prices, many subprime mortgage owners defaulted as they could not meet payments. The increasing defaults of the subprime mortgage holders led to considerable losses for ABSs investors. Moreover, the expectation that all the subprime borrowers would have stopped paying contributed to the drop in ABSs issuance, which dramatically fell in 2007 and 2008. The consequent drop in ABSs prices did not involve only the speculative grade securities, but also those more highly rated (that are those with the protection of the underlying tranches, which are called to cover the losses of the first defaults). In this latter case, it is difficult to think that there was the default as the only dominant factor; rather, the price collapse was linked to a liquidity crisis in addition to the downgrading risk. The drop in demand of mortgage-backed securities in other words caused the sudden illiquidity of the market. This contributed to the spread of the financial crisis together with the scarce consideration of the credit risk related to the securities resulting from securitization, so much that their quotations dramatically fell. This is related also to the role played by the credit rating agencies, called to assign an opinion to the issue of structured ABSs. The financial crisis showed the criticality of such players, as they had assigned ratings that were too high to ABSs (Dodd, 2007). Investors often depend on the rating agencies' opinion for assessing the risk of their investments. In those years, investors were too optimistic about the riskiness of ABSs underwritten or they did not properly understand it. In the final analysis, therefore, the financial crisis was worsened by both inadequate risk assessment by credit rating agencies and over-dependence on these ratings by investors.

In summary, the financial crisis showed that securitization, together with a greater volume of credit derivatives (see Box 3.2), contributed to a significant increase of the size of the financial sector, which was accompanied by an increasingly high leverage at a systemic level. All this promoted excessive growth and at the same time created weaknesses that exacerbated the severity of the crisis and its negative impact on the real economy. Actually, such a situation had quite a domino effect influencing the financial system on a global level.

Box 3.2: Credit Derivatives

Credit derivatives are financial instruments which allow the transfer of the credit risk without selling the underlying assets.

There are several types of credit derivatives. Among the most basic schemes, there is the credit default swap, under which the loans owner (interested in receiving a guarantee) undertakes to pay a premium to the seller of the protection against the guarantee of payment upon the occurrence of a future and uncertain credit event (such as the downgrading, failure to pay, repudiation, etc.) showing the downgrading of the credit standing of a certain counterparty (see for further details Mengle, 2007).

Among the financial intermediaries most impacted by the crisis, there were banks with a great incidence of mortgage loans on their books; that was the case, for example, of the Northern Rock from the UK (Chick, 2008), characterized by a strong mismatching between the maturity of such assets and the short-term maturity of the liabilities (see Box 3.3). Additionally, the impact was significant for banks which had heavily invested in structured securities (among others, Bearn Stearns from the USA and UBS from Switzerland—Bamber, 2009) or had contracts requiring them to support SPVs.

Generally, for many banks the assets became illiquid for the implosion of the underlying market (subprime mortgages market) and consequently they were almost unable to (quickly) evaluate their risk exposures. As is evident, the criteria for evaluating assets at their market-value made banks particularly vulnerable to the price decrease. In such a context indeed banks had losses, which worsened their funding liquidity conditions, pressing them to shed more assets; in this way a spiral process of price decrease and losses increase was activated (Hellwig, 2009; Ryan, 2008).

The general approach of the international authorities in that period was to offer liquidity facilities in order to avoid a long-term period of tight liquidity for banks. Quickly, the scenario deeply changed and, as is well known, on 15 September 2008 the bankruptcy of Lehman Brothers (at that time, the fourth US largest bank) was announced. It is worth mentioning that firstly the US authorities tried to persuade other banks to take over the Lehman Brothers, without however completing the operation successfully. Hence, the decision to allow the bank to fail. All this threw financial markets in panic considering that banks like Lehman Brothers were not too big (and relevant) to fail, as mistakenly believed until then (Labonte, 2014). So, it might be the same for other banks, especially in the light of the exposure they could have on Lehman Brothers, which rapidly became worthless.[3] The interconnectedness among several intermediaries, especially if systemically relevant, was the main source of the instability of the financial system at a global level: from the US to Europe and from a financial intermediary to others all over the world, regardless of the country and the financial institution (even if the big ones played a key-role). Moreover, tensions extended from the financial sector to the real economy, resulting in the Great recession also in the euro area (Strahan, 2013).

Focusing on Europe, it is worth underling that the ECB reacted promptly following various routes (ECB, 2009). First, it provided measures to meet the banking sector's high liquidity needs. The aim

[3] Among others, let us recall the case of one of the largest US money market funds (Reserve primary fund), which owned USD 700 million of short-term paper issued by Lehman Brothers. After the collapse of Lehman, the fund faced great financial difficulties with negative effects for all the money market funds.

was to compensate the interbank market turmoil, where the only operations were characterized by very short-term maturities and/or very high spreads. Additionally, the ECB provided credit to banks at a fixed interest rate and banks were allowed to use a greater range and quota of their assets as collateral in the financing from the central bank. The final goal was to support banks and limit the risk of a prolonged situation of illiquidity of the market, which may cause several distortions like the bank runs (see the already mentioned case of the Northern Rock—Box 3.3), and ultimately to avoid more severe financial distress. To this end, European authorities, equal to the USA, have been forced to intervene with rescue actions in order to mitigate the impact of a systemic meltdown. Support was given to European banks under various forms and in different countries (see Box 3.4) by using public funds. In this latter regard a debate emerged, especially over the trade-off between the final objective to reach the financial stability and the costs incurred by governments and taxpayers.

Box 3.3: The illiquidity issue: The Northern Rock as an Example

At end-2006 the Northern Rock (NR) balance sheet had strongly grown to 101 billion £ (at end-1997 it was 15.8 billion £). The largest portion of assets was represented by residential mortgages. On the liability side, the main source of funding was constituted by wholesale funding (i.e., professional investors); it was about 25% of the total funding and its maturity was almost short (50% had a short-term maturity).

The critical issue was linked to the structure of the NR balance sheet, as the high leverage together with the mismatching of the maturity structure of assets and liabilities created the conditions of great vulnerability with consequent panic on the market and bank runs.

In such a context, it is important to mention the role played by the Bank of England. At first, the Bank of England opposed any support. The initial reason was that the Bank of England believed that such a support could have generated risk-taking and opportunistic behaviour by other intermediaries, pointing to the risk of moral hazard. But, on 14th September 2007, NR asked for and received a liquidity support facility from the Bank of England.

However, the bank runs followed. The causes of the runs were essentially the market fear about both the NR going concern and the possibility that deposits (especially those above a certain threshold) were not fully guaranteed. It was unavoidable for the Bank of England to decide to guarantee the NR deposits.

On 22th February 2008 the NR was nationalized after two unsuccessful attempts of private takeovers (see for further details Shin, 2009).

Box 3.4: Banks Support by European Authorities (Some Examples)
Benelux (Belgique, Netherlands, Luxembourg). The financial company Fortis was bailed out by the Benelux authorities.

Netherlands. The government nationalized Fortis Dutch activities. The other activities were sold to BNP Paribas (France).

France, Belgique and Luxembourg. The authorities provided support for the re-capitalization of the financial conglomerate Dexia.

UK. Authorities took over Bradford and Bingley (operating in the real estate market) mortgages and loans, while savings operations and branches were sold to the Spanish Santander Bank. Besides the authorities participated in equity of two large banks: Royal Bank of Scotland and Lloyds Bank.

Germany. Hypo Real Estate obtained a facility from a group of German banks and the German government

The decisions carried out by the European authorities during the early steps of the financial crisis were not fully coordinated, rather they were more related to the peculiarities, in terms of financial difficulties, met by the single financial intermediary and/or country. Over time, it has become increasingly clear the need to refer to structural and coordinated measures, in order to ensure, among others, the stability of the euro area. Among such measures, as mentioned, the creation of the Banking Union represents one of the key-initiatives (see next Sections).

3.3 The Banking and Financial Integration Process: Main Steps

The process of financial integration within Europe has taken place gradually thanks to various steps. Below, we report some of the most relevant for the banking system and the financial markets presenting them in chronological order.

First, the Directive 77/780/EEC of 12 December 1977 on the coordination of the laws, regulations and administrative provisions relating to the taking up and pursuit of the business of credit institutions (CEC, 1977). This Directive (also named First Banking Directive) aimed at contributing to the start of the process of harmonization throughout the European banking system, which over the recent years has been strongly enhanced, as it focused on the need for common standards, particularly on the most relevant matters (solvency, liquidity, internal control, etc.). Even if such ambitious targets have required longer time and other measures, it is worth noting the importance of the starting point of a process intended to culminate in the European Banking Union. In the meantime, each national banking system continued to adopt specific basic prudential regulations (capital adequacy framework, see Sect. 3.4.1).

Another step is represented by the Directive 88/361/EEC of 24 June 1988 on the liberalization of capital flows within EU, entered into force on 1 July 1990 (CEC, 1988). The objective was to progressively eliminate the constraints on free capital movements; controls were limited to speculative cases.

In 1989, the Second Banking Directive had been issued: Directive 89/646/EEC of 15 December 1989 on the coordination of laws, regulations and administrative provision relating to the taking up and pursuit of the business of credit institutions (CEC, 1989). Amending the First Banking Directive, this one represented the real prerequisite for the launch of the European single market in the financial sector. To this end, it allowed banks of the EU member States to provide activities subject to mutual recognition throughout the EU territory, on the basis of the licensing obtained from the home country authorities. This is the so-called Home Country Control, when the home country authorities are responsible for supervision on solvency for foreign and national subsidiaries of a bank. Under the Second Banking Directive therefore the mutual recognition principle had been adopted and the financial, banking and insurance activities (products and services) subject to it are those that could be performed (also in another member State) by any financial institution with a headquarter in one of the EU member States. Such activities could be provided without the licensing from the host country authorities. It is possible to conclude that such a principle had given birth to a competitiveness among several banking regulations.

In the late 90s, the Financial Services Action Plan had been issued (1999) by the European Commission in order to eliminate the remaining regulatory and legal barriers restricting the provision of financial services and the free movement of capital across the EU (CEC, 1999). Additionally, it was intended to contribute to a level playing field among all the market players. The Action Plan particularly intends to accomplish the following purposes: the integration of the wholesale financial markets; the promotion of retail markets and services in order to expand opportunities for consumers; and the harmonization and enhancement of the supervisory framework (see next Section).

In 2012, the decision of creating the European Banking Union was adopted by the European authorities (ECB, 2012). Particularly in June 2012, the European Council underlined the need of carrying out a set of structural reforms intended to address the impacts of the international crisis (see Sect. 3.2). All the measures aimed at promoting the economic integration among the EU member States, including the creation of the Banking Union and the introduction of a common regulatory framework for all the EU banks. This initiative, as mentioned, resulted from the need for the EU of decisively responding to the economic and financial crisis. Besides, the global crisis had highlighted that supervision at a national level was not the best viable solution, especially in light of certain factors, which across recent years have become increasingly important for the stability of the EU banking system. They are the presence of big financial institutions, the high interconnectedness among them and the link between banks and national governments. Evidence on the distress of cross-border banks in Europe pointed out that during the crisis national measures were mainly focused on maintaining domestic peculiarities, without fully considering the integrated value of banks.

To conclude, the transition from a national supervisory system, where national authorities were responsible for guaranteeing the domestic banking system stability, to a centralized supervisory system, within the banking union, seemed to represent a good arrangement for the soundness of the financial system at EU level, as well as for breaking the loop between banks and sovereigns.

Another important step is the Capital Markets Union (CMU) project presented for the first time in 2015 (and subsequently revised in 2017 and in 2020 in order to consider the pandemic crisis) by the European Commission (EC, 2015a; 2017; 2020). It is an Action Plan aimed at promoting the development of capital markets across Europe, by intervening in particular on the diversification of the funding sources for corporates (particularly European Small and Medium Enterprises,

SMEs, for which funding is mainly dependent on the banking channel), the increase of the opportunities for investors and the full achievement of the capital flows. The functioning of an integrated capital market could definitely contribute to the growth of the European economy as a whole, following a sustainable approach and aiming to increase its competitiveness at a global scale.

The CMU project is considered relevant for the achievement of important goals at a European level: the post-pandemic crisis recovery; the creation of a resilient and inclusive economy; and the transition towards a digital and sustainable economy. All these goals require significant investments, which cannot be dealt with as the only means of public resources and bank lending. Rather, they need the key-contribution of the integrated capital markets as the only way to face the recovery and transition processes.

The pandemic crisis made the CMU even more urgent. As is well known during the crisis induced by COVID-19 corporates and households were largely supported by the governments and banks for their liquidity needs. The effectiveness of the response to the emergency situation does not solve the problem in the medium-long term. Indeed, European firms, especially SMEs, need financial resources in order to strongly counter the effects of the various shocks (financial, pandemic and war crisis) and ultimately becoming more resilient.

As above mentioned, the CMU has been reviewed a first time in 2017 when the European Commission has established some measures, partly included in the previous version, in order to meet three objectives: (i) supporting the sustainable, digital and resilient recovery of the economy, making more accessible the funding for European firms; (ii) creating a safer European market for savings and investments; and (iii) integrating the national capital markets into a single one (EC, 2017).

As regards the first goal, it is worth underling, though fairly obvious, that the ability of the post-pandemic crisis recovery is closely connected with the availability of financial resources by firms. European banking lending will contribute, but firms will particularly need equity or more stable liabilities (medium-long-term debt). Over the years, financing through stocks and bonds played a complementary role to banking lending, even if with a different degree from country to country. The CMU is aimed at further enhancing such complementarity. To this end, there is the priority of promoting the functioning of direct financing, by balancing the investors needs of receiving reliable information and the firms ones of limiting the expenses related to the disclosure to the market. The availability of reliable information for investors and all the other stakeholders is important also for favouring the accessibility of data on sustainability with the effect, ceteris paribus, of greater sustainable investments. On the other hand, direct financing could be expensive and complex for SMEs. The simplification of rules could contribute to the decrease of the administrative, listing and compliance costs and therefore eliminate one important barrier which precludes SMEs to the capital markets.

With regard to the second objective, it is worth noting that despite the high savings rate in Europe; European investors participation to the capital markets is quite modest. This is mainly due to the low comparability of similar investment products that are regulated by different legislation and consequently subject to different disclosure requirements. It is clear that this makes informed investment decisions more difficult. Additionally, informed investment decisions need to be taken based on specific knowledge and expertise; in other words the financial competences are a key-factor for understanding the real functioning of the market. It is not so obvious for retail investors.

The scarce participation of retail investors to capital markets represents a barrier for firms in, among others, collecting medium-long-term funding, thus penalizing the recovery of the economy under the sign of sustainability and digitalization. On the other hand, such a barrier does not allow investors to benefit from opportunities in terms of appropriate returns. Hence, the CMU aims at revitalizing the investment opportunities for retail investors, also considering appropriate risk-return mixes, as well as contributing to the restoration of confidence in the capital markets, which in turn could result in the increase of medium-long-term investments.

The third goal of CMU wishes to recognize to the European capital markets a higher relevance in line with the importance that the European economy has. Currently, the European capital markets are still too fragmented among the different legislation with negative effects in terms of inability to attract a wide investor base, as well as of a competitive disadvantage for the financial institutions compared to their global peers.

In conclusion, the CMU, as a medium-long-term project, requires big efforts based on the paradigm of banking system-capital markets complementarity as well as the reduction of regulatory, cultural and territorial barriers.

The 2020 review of the CMU goes on in this direction, by restating the key-contents (green and digital transition, a more inclusive economy, etc.), and underlying certain needs closely linked to the pandemic crisis (EC, 2020).

3.4 The Rational of Banking Supervision: Some Key-Points

As above-mentioned, the need of ensuring the financial stability also for the benefit of the real economy within the euro area forced the European authorities to take the decision of creating the Union Banking. Particularly, the fear that a further spreading of the sovereign debt crisis could make the situation resulting from the international financial crisis worse became the prerequisite for such an initiative. To this end in June 2012, the euro area governments decided to assign the supervisory powers to the ECB within the Single Supervision Mechanism (SSM) framework, which entered into force in November 2014 (EC, 2012). The SSM stated the movement of the supervision on banks from national supervisors to ECB by applying a uniform approach to supervision thanks to the harmonization of practices and methodologies. This ultimately helps comparability across the euro area countries and contributes to the limitation of compliance costs for banks. Additionally, a Single Resolution Mechanism (SRM, since 2016) has been created, aimed at orderly managing the distressed financial institutions (EC, 2015b).

Such disposals are based on a set of harmonized laws (Single Rulebook) that all EU banks must comply with. They state rules on the financial sector in all EU countries, such as capital adequacy requirements, recovery and resolution plans and system of harmonized national Deposit Guarantee Schemes.

Supervision plays a crucial role in ensuring the sound and prudent management of banks, as well as the stability, the competitiveness and effectiveness of the financial system as a whole. A financial institution authorized and supervised in its home country has the opportunity to expand its business across the EU (i.e. through branches) without additional supervision.

Supervision is a process which could have various forms. One relates to the first entry of the financial institution into business. In order to release the corresponding authorization, supervisory authorities have to assess qualifications, expertise, integrity and honesty of people involved, in order to decide on their attitude to manage the bank. The assessment concerns also the ownership and governance structure of the intermediary, with particular reference to the boards of directors, internal controls system, risk management and level of capitalization.

Another form characterizes the stage of the ongoing banking business, when supervision is particularly focused on the soundness of the institutions and the financial system. In this case, the analysis refers to the asset quality, capital adequacy, liquidity, etc., and the tools used are various: inspection, audit and data collecting and processing by authorities. Generally, in case of non-compliance with laws and regulations as well as in presence of bad management and/or frauds, penalties and sanctions are imposed to banks by supervisors.

Lastly, authorities are called to intervene in the case of crisis management in order to mitigate the effects of a certain financial institution distress and its spread to the rest of the financial system (systemic risk).

3.4.1 Focus on Capital Adequacy Framework

One of the key-topics in banking supervision is the capital adequacy framework, aimed at ensuring that a bank holds an appropriate amount of capital, considering the risks connected to the business performed.

As is well known, capital performs as a foundation for the growth of any business and as a buffer against unexpected losses. In the case of banks, it plays a crucial role, as only adequately capitalized banks are able to face losses and at the same time continue to grant loans to the real economy (households and firms) regardless the business cycle. This is particularly true during negative economic cycles: only banks with an adequate level of capitalization are able to provide credit to the market despite the adverse environment. Additionally, appropriate levels of capital could improve market confidence in the banking system, as it is perceived more stable and therefore more reliable. From this derives the need, from both the side of the single bank and supervisors, to determine which is the adequate capital amount able to guarantee the protection against unexpected losses. Too low capital prevents banks to face losses, too high capital could undermine the most efficient use of financial resources and therefore negatively affect the lending capacity. To this end, the international supervisory authorities have introduced specific regulation on capital adequacy for banks, meaning the need to determine a measure of capital (minimum capital requirement) able to ensure efficiency and stability to the financial system, also by lowering the risk of bank's default. In this sense, it is worth noting that banks capital adequacy is relevant also to mitigate the risk of a domino effect (see Sect. 3.2) that occurs when the distress of a bank, or the fear of such a failure, spreads to the rest of the financial system (systemic risk).

The first international disposal on banking capital adequacy was introduced in 1988 by the Basel Committee on Banking Supervision (BCBS)[4] (BCBS, 1988). The decision was particularly based on the fear that the early 80s Latin American debt crisis could amplify the instability of the international markets. In such a scenario, it was deemed appropriate to avoid the deterioration of banks capital all over the world, by adopting a harmonized approach for the capital adequacy measurement. This disposal was indeed considered good for enhancing the stability of the international banking system and to remove differences in national capital adequacy frameworks. All this resulted in the first Basel Capital Accord (known as Basel I), focused on the compliance by banks with a minimum (8%) capital to risk-weighted-assets (RWAs) requirement (Regulatory Capital/RWAs $\geq$ 8%). Under Basel I, such weighted risk ratio was aimed at assessing capital in relation only to the credit risk, as the main risk related to the banking core business and specifically related to the possibility that a bank should face a loss due to the default and/or downgrading of the borrowers (households and firms). The weights were defined in a standardized way that is on the basis of predetermined (by supervisors) percentages (from 0 to 100%) dependent on the nature of the counterparty (e.g., central governments, financial intermediaries, firms, etc.), the possible presence of guarantees as well as the country risk. The Basel I framework represented a very important first step in the process of capital adequacy regulation, as it allowed banks to definitely increase capitalization and enhance the link between the risk exposure and taking decisions, especially with reference to capital management. At

[4] The BCBS was set up in 1974 by the Governors of the Central Banks of the G10 following turbulences in international currency and banking markets. It is the main global standard setter for banking regulation. Its aim is to enhance regulation and supervision on a global scale in order to ensure financial stability. Its decisions do not have legal force. Once, it defines the regulation standard; these should be brought into force by the BCBS members under the specific rule-making process, which could decide for stricter rules.

the same time, such a framework had some shortcomings, as the exclusive treatment of the credit risk without considering the other types of banking risks.

This was one of the reasons of the evolutionary process undertaken by supervisors during the following years, aimed at progressively improving the framework considering the changing environment where banks at a global level should perform their business. In this respect, it is worth mentioning the inclusion in the mid-90s of the risk of loss in balance sheet positions resulting from changes in market prices (so-called market risk) into prudential discipline, so that the RWAs also considered the market risk in addition to the credit risk.

More in general, the need of further improving the correlation between capital absorption and risks undertaken by banks together with the growing improvement in risk management activities by banks led to reforming the Basel I framework. In the mid-2000s, this took the form of Basel II, built upon three pillars: definition of the minimum capital requirements; supervisory review; market disclosure (BCBS, 2004).

The first pillar extended the calculation of the RWAs to the operational risk that is the risk of losses from inadequacy or dysfunction of procedures, human resources and internal systems, as well as exogenous events. It also provided for alternative approaches for the quantification of capital requirements and characterized by different degrees of complexity (not only standardized methods but also internal data based systems approaches). Ultimately, banks had to be compliant with the capital to RWAs ratio that remained at least 8%, but the new framework, as mentioned, under the first pillar introduced a wider range of risks (credit, market and operational risks) and alternative methodologies for the quantification of the RWAs. The Basel II second pillar introduced an evaluation process by banks on their own capital adequacy in relation to the total risk exposure as well as the verification procedure by supervisors, called to formulate an overall judgement on the bank and eventually impose corrective measures. The third pillar was focused on the obligation for banks to disclose information on capital adequacy, the total risk exposure and the control systems to the market.

The evolutionary process continued after the 2007–2008 international financial crisis. The need to review the prudential regulation in order to ensure stability to the financial system considered as a whole gave birth to Basel III, still in force (BCBS, 2017). To this end, conserving the conceptual structure of the previous framework, Basel III is articulated into three pillars and contains additional prudential measures that banks should comply with. Among these, under the first pillar in particular, Basel III introduces higher capital requirements, a leverage ratio aimed at limiting the growth of debt and liquidity risk standards. In other words, Basel III is a more comprehensive framework of prudential measures compared to the previous ones. Its overall purpose is to strengthen banks' ability to face adverse scenarios, thus mitigating the risk of spillover from the financial sector to the economy. At the same time, it forces banks to improve their governance and risk management as well as to enhance their disclosure to the market. Another characteristic of Basel III is the fact that it has introduced bank-level prudential measures (micro-prudential approach) intended to increase the resilience of each financial institution during stress periods. Besides, it has included also a macro-prudential approach on which basis attention is given to the stability of the financial system considered as a whole. Only, the interaction of the two approaches is able to ensure greater resilience to the single bank and mitigate the risk of system-wide shocks.

The refining process is still ongoing, in order to further improve the capital adequacy framework considering the various crisis occurred since the 2007–2008 (financial, sovereign, pandemic and war crisis) and the consequent need to ensure that the banking system could concretely and continuously contribute to the financial support to the real economy.

To conclude, the underlying logic of the succession of the revision of regulation on capital adequacy over the last decades is the need to ensure the stability of the financial and economic systems at global level considering the possible threats that could derive from adverse scenarios and the consequent negative impacts they could have on the economies.

References

Acharya, V. V., & Richardson, M. (2009). Causes of the financial crisis. *Critical Review, 21*(2–3), 195–210. https://doi.org/10.1080/08913810902952903

Allen, F., & Carletti, E. (2010). An overview of the crisis: Causes, consequences, and solutions. *International Review of Finance, 10*(1), 1–26. https://doi.org/10.1111/j.1468-2443.2009.01103.x

Bamber, B. (2009). *Bear trap, the fall of bear stearns and the panic of 2008 (HC)*. ibooks. https://www.ibs.it/bear-trap-fall-of-bear-libro-inglese-bill-bamber-andrew-spencer/e/9781899694853

Basel Committee on Banking Supervision—BCBS. (1988). *International convergence of capital measurement and capital standards*. https://www.bis.org/publ/bcbs04a.htm

Basel Committee on Banking Supervision—BCBS. (2004). *Basel II: International convergence of capital measurement and capital standards: A revised framework*. https://www.bis.org/publ/bcbs107.htm

Basel Committee on Banking Supervision—BCBS. (2017). *Basel III: Finalising post-crisis reforms*. https://www.bis.org/bcbs/publ/d424.htm

Bernanke, B. S. (2008). *Financial markets, the economic outlook, and monetary policy*. https://ideas.repec.org/p/fip/fedgsq/352.html

Blundell-Wignall, A. (2007). The private equity boom: Causes and policy issues. *Financial Market Trends, 2007*(1), 59–86. https://doi.org/10.1787/fmt-v2007-art4-en

Blundell-Wignall, A., Atkinson, P., & Lee, S. H. (2009). The current financial crisis: Causes and policy issues. *OECD Journal: Financial Market Trends*, *2008*(2), 1–21. https://doi.org/10.1787/fmt-v2008-art10-en

Chick, V. (2008). Could the crisis at Northern Rock have been predicted?: An evolutionary approach. *Contributions to Political Economy*, *27*(1), 115–124. http://www.storep.org/workshopsiena/Chick2.pdf

Colloquium, A., & Papademos, L. (2010). *Lessons for financial stability and monetary policy*. https://www.ecb.europa.eu/pub/pdf/other/greatfinancialcrisisecbcolloquiumpapademos201206en.pdf

Council of the European Communities—CEC. (1977). *First Council Directive 77/780/EEC of 12 December 1977 on the coordination of the laws, regulations and administrative provisions relating to the taking up and pursuit of the business of credit institutions*. https://eur-lex.europa.eu/legal-content/EN/TXT/?uri=CELEX%3A31977L0780

Council of the European Communities—CEC (1988). *Council Directive 88/361/EEC of 24 June 1988 for the implementation of Article 67 of the Treaty*. https://eur-lex.europa.eu/legal-content/EN/ALL/?uri=CELEX:31988L0361

Council of the European Communities—CEC (1989). *Second Council Directive 89/646/EEC of 15 December 1989 on the coordination of laws, regulations and administrative provisions relating to the taking up and pursuit of the business of credit institutions and amending Directive 77/780/EEC*. https://eur-lex.europa.eu/legal-content/EN/TXT/?uri=CELEX%3A31989L0646

Commission of the European Communities—CEC (1999). *Implementing the framework for financial markets: Action plan*. https://eur-lex.europa.eu/legal-content/EN/TXT/PDF/?uri=CELEX:51999DC0232&from=nl

Cukierman, A. (2013). Monetary policy and institutions before, during, and after the global financial crisis. *Journal of Financial Stability*, *9*(3), 373–384. https://doi.org/10.1016/j.jfs.2013.02.002

Deku, S. Y., Kara, A., Deku, S. Y., & Kara, A. (2017). The Role of Securitization in the 2007–2009 Crisis. *Securitization: Past, Present and Future*, 131–143. https://link.springer.com/chapter/10.1007/978-3-319-60128-1_7

DiMartino, D., & Duca, J. V. (2007). The rise and fall of subprime mortgages. *Economic Letter, 2*. https://ideas.repec.org/a/fip/feddel/y2007inovnv.2no.11.html

Dodd, R. (2007, December). Subprime: Tentacles of a crisis. *Finance & Development*. Available at https://www.imf.org/external/pubs/ft/fandd/2007/12/dodd.htm

Duca, J. V., Muellbauer, J., & Murphy, A. (2010). Housing markets and the financial crisis of 2007–2009: Lessons for the future. *Journal of Financial Stability, 6*(4), 203–217. https://doi.org/10.1016/j.jfs.2010.05.002

European Central Bank—ECB (2007). *How the Euro became our money: A short history of the Euro banknotes and coins*. https://www.ecb.europa.eu/pub/pdf/other/euro_became_our_moneyen.pdf

European Central Bank—ECB. (2009). *The financial crisis and the response of the ECB*. https://www.ecb.europa.eu/press/key/date/2009/html/sp090612.en.html

European Central Bank—ECB (2012). *Towards a European Banking Union*. https://www.ecb.europa.eu/press/key/date/2012/html/sp120907.en.html

European Commission—EC. (1985). *Completing the Internal Market: White Paper from the Commission to the European Council (Milan, 28–29 June 1985)*. https://op.europa.eu/en/publication-detail/-/publication/4ff490f3-dbb6-4331-a2ea-a3ca59f974a8/language-en

European Commission—EC (2012). *A roadmap towards a Banking Union*. https://eur-lex.europa.eu/legal-content/EN/TXT/?uri=CELEX%3A52012DC0510

European Commission—EC (2015a). *Action plan on building a capital markets union*. https://eur-lex.europa.eu/legal-content/EN/TXT/?uri=CELEX%3A52015DC0468

European Commission—EC. (2015b). *Single resolution mechanism to come into effect for the Banking Union*. https://ec.europa.eu/commission/presscorner/detail/en/IP_15_6397

European Commission—EC. (2017). *Mid-term review of the capital markets union action plan*. https://finance.ec.europa.eu/publications/mid-term-review-capital-markets-union-action-plan_en

European Commission—EC. (2020). *Capital Markets Union: Commission to boost Europe's capital markets*. https://finance.ec.europa.eu/capital-markets-union-and-financial-markets/capital-markets-union/capital-markets-union-2020-action-plan_en

European Commission—EC. (2022a). *The European Union—What it is and what it does*. https://op.europa.eu/webpub/com/eu-what-it-is/en/#chapter13

European Commission—EC. (2022b). *The Euro area: summary*. https://data.europa.eu/doi/10.2765/832612

European Court of Auditors—ECA. (2020). *Capital markets union. Slow start towards an ambitious goal*. Special report No 25, 2020. https://op.europa.eu/en/publication-detail/-/publication/ec3b0d43-61e5-11eb-aeb5-01aa75ed71a1/language-en/format-PDF/source-280285451

European Parliament—EP (1992). *The treaty of European Union: The position of the European Parliament*. https://op.europa.eu/en/publication-detail/-/publication/3d8f418f-08f2-4afc-9cda-a9ae97099450/language-en/format-PDF/source-280241390

European Parliament—EP (1999). *Citizenship of the Union: possibilities, recommendations and suggestions for protecting and extending citizenship of the Union with a view to the 1996 Intergovernmental Conference to review the Maastricht Treaty*. https://op.europa.eu/en/publication-detail/-/publication/c1060f94-f3c4-4aef-890d-f32413bc89d0/language-en/format-PDF/source-280241770

Glockner, I., & Rittberger, B. (2012). The European Coal and Steel Community (ECSC) and European Defence Community (EDC) treaties. In *Designing the European Union: From Paris to Lisbon* (pp. 16–47). https://link.springer.com/chapter/10.1057/9780230367579_2

Goodhart, C. A. (2008). The background to the 2007 financial crisis. *International Economics and Economic Policy*, *4*(4), 331–346. https://link.springer.com/article/10.1007/s10368-007-0098-0

Gorton, G., & Metrick, A. (2013). Securitization. In *Handbook of the economics of finance* (Vol. 2, pp. 1–70). Elsevier. https://www.nber.org/system/files/working_papers/w18611/w18611.pdf

Gortsos, C. V. (2017). A brief overview of the European Banking Union. *L'Europe en formation* (2), 61–83. Available at https://www.cairn.info/revue-l-europe-en-formation-2017-2-page-61.htm

Healey, N. M. (1995). From the treaty of Rome to Maastricht: The theory and practice of European integration. In *The economics of the New Europe* (pp. 17–58). Routledge. https://www.taylorfrancis.com/chapters/edit/10.4324/9780203427309-4/treaty-rome-maastricht-nigel-healey

Hellwig, M. F. (2009). Systemic risk in the financial sector: An analysis of the subprime-mortgage financial crisis. *De economist*, *157*, 129–207. https://link.springer.com/article/10.1007/s10645-009-9110-0

Labonte, M. (2014). *Systemically important or "too big to fail" financial institutions*. https://ecommons.cornell.edu/handle/1813/77638

Lucas, D. J., Goodman, L. S., & Fabozzi, F. J. (2006). *Collateralized debt obligations: Structures and analysis* (Vol. 140). Wiley.
McKibbin, W. J., & Stoeckel, A. (2010). The global financial crisis: Causes and consequences. *Asian Economic Papers, 9*(1), 54–86. https://doi.org/10.1162/asep.2010.9.1.54
Mengle, D. (2007). Credit derivatives: An overview. *Economic Review-Federal Reserve Bank of Atlanta, 92*(4), 1. https://citeseerx.ist.psu.edu/document?repid=rep1&type=pdf&doi=70bf8e0889e3eb3854b23d3a020cef2eaa223fc4
Moutot, P., Jung, A., & Mongelli, F. P. (2008). *The working of the Eurosystem: monetary policy preparations and decision-making-selected issues* (ECB Occasional Paper 79). https://www.ecb.europa.eu/pub/pdf/scpops/ecbocp79.pdf
Myers, S. C., & Rajan, R. G. (1998). The paradox of liquidity. *The Quarterly Journal of Economics, 113*(3), 733–771. https://doi.org/10.1162/003355398555739
Phinnemore, D. (2013). The European Union: Establishment and development. *European Union Politics*, 26–40.
Phinnemore, D., & Warleigh-Lack, A. (Eds.). (2009). *Reflections on European integration: 50 years of the treaty of Rome*. Springer.
Quaglia, L., Howarth, D., & Liebe, M. (2016). The political economy of European capital markets union. *Journal of Common Market Studies, 54*, 185.
Ramskogler, P. (2014). Origins of the crisis–drawing the big picture. *NAEC monograph, OECD, Paris*. https://www.oecd.org/naec/naec_origins-of-the-crisis_eng.pdf
Risenfeld, S. A. (1992). The coming of Europe. *Hastings International and Comparative Law Review, 16*, 461.
Ryan, S. G. (2008). Accounting in and for the subprime crisis. *The Accounting Review, 83*(6), 1605–1638. https://doi.org/10.2308/accr.2008.83.6.1605
Shin, H. S. (2009). Reflections on Northern Rock: The bank run that heralded the global financial crisis. *Journal of economic perspectives, 23*(1), 101–119. https://pubs.aeaweb.org/doi/pdf/10.1257/jep.23.1.101
Solomon, D. (2012). The rise of a giant: Securitization and the global financial crisis. *American Business Law Journal, 49*, 859.
Strahan, P. E. (2013). Too big to fail: Causes, consequences, and policy responses. *Annual Review of Financial Economics, 5*(1), 43–61. https://www.annualreviews.org/doi/abs/10.1146/annurev-financial-110112-121025
Taureck, R. (2006). Securitization theory and securitization studies. *Journal of International Relations and Development, 9*(1). https://core.ac.uk/download/pdf/48444.pdf

Véron, N. (2013). *A realistic bridge towards European banking union* (No. 2013/09). Bruegel Policy Contribution. https://www.econstor.eu/bitstream/10419/106306/1/751511579.pdf

Véron, N., & Wolff, G. B. (2016). Capital Markets Union: A vision for the long term. *Journal of Financial Regulation, 2*(1), 130–153. https://doi.org/10.1093/jfr/fjw006

CHAPTER 4

Banking Support for Firms: Some Business Areas

Abstract This chapter illustrates some of the possible banking business areas in favour of firms. To this end, a focus on the traditional lending activity is given as well as on other financial services a bank may provide to firms. The underlying logic is to understand the extension of the offer of financial products, which may promote the enhancing of the banks-firms relationship along a perspective of an increasing integration between financial markets and intermediaries.

Keywords Banking business areas · Lending · Commercial banks · Investment banking

4.1 Banks-Firms Relationship: Introductory Remarks

Over the years, the several changes and the evolution in the global financial markets as well as the impulses to diversify the sources of funding, especially for the small and medium enterprises—SMEs, in order to promote a closer complementarity between direct and indirect financing (among others we mention the Capital Markets Union, see Chapter 3) have progressively driven to an increasing financial markets-financial intermediaries integration all over the world (IMF, 2022).

P. Ferretti and P. Martino, *Banking and Financial Markets*,
https://doi.org/10.1007/978-3-031-32562-5_4

Even if with different intensity due to the specificities of the single financial system, this has led to a general development of more integrated relationships between firms and banks where the latter offer not only lending, but also a wide range of more sophisticated financial services according to a relationship banking approach. Hence, the entry into new business areas (Sect. 4.3) has increasingly become an opportunity to be grasped by banks, in order to continue to play a relevant role in the financial support to firms and in the meantime ensure the protection of their budget balances, put to the test by the growing competitiveness on an international scale and, during more recent years, by the multiple (financial, pandemic, war) global crises (Howcroft et al., 2007; Montedoro, 2020; Song et al., 2018).

In other words, the ability of banks to propose a wider offer of financial products and services to corporates has been a crucial driver for strengthening bank-firm relationships (IMF, 2022). In such a context, the supply of credit has continued to be a key-component in bank-firm relations, but it has been included into a more complex offer in the name of the above-mentioned relationship banking approach, based on the high intensity of the relations and the provision of not only credit products to enterprises. In hindsight, on the basis of this logic, the supply of credit has evolved, increasingly becoming more innovative and flexible, consistent with the changed operational environmental. To this end, particularly after the financial crisis, banks have progressively focused on more accurate analysis of the features of corporate financial needs, granting loans with conditions and terms consistent with the characteristics of the funded businesses (OECD, 2013, 2019). This asks banks also to implement adequate management procedures and processes aimed at promoting effective mechanisms of loans origination and monitoring as well as best methodologies for the assessment of the related credit risk. This is an approach that tends to favour a medium-long-term perspective, focused on the in-depth analysis of the firms' hallmarks, as a prerequisite for the appropriate evaluation of their ability to create value in the future (EBA, 2020; ECB, 2017).

On the basis of the previous considerations, it is possible to identify some business areas that allow the profile of the mentioned corporate banking approach to be defined. In addition to the lending activity, as core business of the commercial banks in particular, it is possible to include investment banking and other activities (Sects. 4.3 and 4.4).

It is pointless underlying that a wider banking offer needs also organizational changes in favour of innovation, training and talent management, as well as a business model in order to strengthen the competitive advantages that could result from business diversification and therefore the enlargement from activities almost exclusively focused on lending towards more added-value and advisory activities (BIS, 2018).

4.2 Bank Lending

As is well-known, the lending activity is the core business of banks, especially commercial ones (ECB, 2020a). This is due to their facilitating role in the transactions between surplus (firms and households with more money than they need) and deficit spending units (firms and households with less money than they need). In other words, they stand between the two types of units and they transform financial claims in order to attract both of them. The goal is to tailor financial products and services aimed to meet corporates' (and households') needs. Banks' financial support to firms is particularly relevant for SMEs, as they do not have easy access to direct financing (OECD, 2018).

Through the lending activity banks are aimed at achieving, particularly in a medium-long-term perspective, profitability targets, as lending is characterized by a quite good risk-return mix (BIS, 2013; Mergaerts & Vennet, 2016). Besides, lending is typically the prerequisite for the strengthening of the relation with firms as it allows banks to offer firms other financial services. In order to improve their market position and integration into the economic environment, banks are hence called to adopt consolidation strategies in the relations with firms. This is relevant also for the achievement of the profitability as well as value creation related to the lending.

The logics underlying the lending activity carried out by banks are linked both to the volume (size) and to the composition of the lending portfolio.

The determination of the size of the lending portfolio depends on several factors, such as the economic cycle, the characteristics of the financial system and lastly endogenous factors.

In general, a positive economic trend is associated, ceteris paribus, with a greater credit supply as a consequence of a greater credit demand by firms resulting from higher financial needs for investments. The opposite is observed for a negative economic trend (ECB, 2013). This occurred recently during the financial and pandemic crisis. In both cases, the economic recession negatively impacted the credit demand, as firms scaled back expansion plans and as a consequence bank lending declined substantially (Ferretti, 2016).

The volume is also affected by the level of development of a certain financial system. More advanced financial systems, ceteris paribus, allow firms to diversify the sources of funding, using the forms of financing (short-term and long-term debt, equity) that are well (best) suited to the characteristics (i.e. maturity, risk profile, etc.) of their financial needs (Vo, 2018).

Lastly, the size of the loans portfolio is connected to the specific features of the financing banks, such as the characteristics of its liabilities; among others indeed it is affected by the qualitative-quantitative composition of the bank's funding (i.e. deposits), which could be able to be an important source of financing of banks' lending activity (ECB, 2017).

On the other side, the decisions connected to the composition of the lending portfolio are driven by diversification strategies applied to the whole portfolio with the aim of decreasing its total risk exposure. Such strategies could be related to amount class, economic sector and geographic area, as well as technical forms of financing (ECB, 2017).

With reference to the amount class, the lending portfolio is built by spreading it on the basis of the size of each loan. In this way, banks are able to limit the variability of the expected revenue of loans by decreasing the probability of large losses, which could severely affect their stability.

Diversification could also refer to the sectors and geographic areas that is when loans are granted to different industries and territories in order to mitigate the risk that a specific economic sector or geographic area may be affected by a crisis and by this way negatively impact on banks stability. In other words, such strategy is aimed at structuring a lending portfolio where debtors are low-correlated in terms of their sectors and territories (Meslier-Crouzille et al., 2015; Shim, 2019).

Lastly, diversification could be achieved by offering different technical forms of loans, that is, characterized by different conditions and terms (i.e. guarantees) able to mitigate the losses resulting from the default (ECB, 2020b).

Moving from the whole portfolio to the single loan, it is worth noting that banks have to decide to finance or not firms on the basis of the assessment of the risk (credit risk) connected to each loan (Naili & Lahrichi, 2022). As is well-known, the credit risk is related to the possibility that a bank should face a loss due to the default and/or downgrading of the borrowers. Banks should assess such a risk gathering and processing information and data on debtors in order to quantify the probability of default and the consequent economic loss. The information to be gathered is of various nature: from the basic data resulting from the funding application to the economic and financial information, evidenced from the balance-sheet or similar reports, to information related to the relations between the applicant and all the lenders as well as information about the competitive positioning of the firm, its management structure, organization, ownership, etc. All the information, adequately processed, is aimed at underlying the risk profile of the firm and, on that basis, deciding the financing and its conditions and terms.

The increasing difficulties connected with the measurement and management of the credit risk, mainly due to the growing complexity and instability of the operational environment on a global scale, have led banks to develop more sophisticated credit risk management methodologies (Doumpos et al., 2019; Van Deventer et al., 2013), currently almost represented by the internal rating based systems (Cucinelli et al., 2018). These may be defined as a set of methods, processes, controls, data collection and IT systems supporting the credit risk assessment, the assignment of the various exposures (transactions or debtors) to ratings and the measurement of the default and loss related to a certain loan. Compared to more traditional approaches, the internal rating systems are focused on more advanced methods of classification and quantification of the risk exposures. Ultimately, they allow banks to identify a graded level of risk associated with each transaction and/or debtor (Ferretti, 2016).

Moreover, it is worth mentioning that the first rating assignment should be continuously monitored by banks during the whole credit relationship, in order to verify the probability of a deterioration of the borrower situation, as in the case of the worsening of the quality of the balance sheet, competitiveness, relation with one of the lenders, etc. In such a case, banks are called to adjust their assessment of the credit risk, and if it is necessary change the conditions and terms previously applied (EBA, 2020).

4.3 Investment Banking

As already mentioned, lending to firms is the core business of commercial banks. The approach they mainly adopt is the relationship one that is based on long-term relations and on an offer including additional banking services.

It may happen that some overlaps between commercial and investment banking are possible. Hence, even if investment banks are mainly focused on the support to firms in selling their security issues (debt and/or equity) in the market as a source of financing (see below); their business may obviously include the lending activity, especially towards large corporates. However, it is more likely that commercial banking is oriented towards smaller corporate customers and that their business is typically provided at regional (and/or national) level, as a result of the closer relations they have with customers. On the other hand, investment banking tends to be characterized by a wider operational scope (Forestieri, 2018).

The Box 4.1. shows a short overview of the evolutionary path of commercial and investment banking.

Box 4.1. Commercial and Investment Banking

In the past years, commercial and investment banking had been separated in an artificial way, by means of legislation. Particularly, it was the Glass-Steagall Act (Banking Act of 1933), the USA legislative initiative, which separated commercial banking from investment banking.

The reasons for that are rooted in the 1929 stock market turmoil and the consequent Great Depression in the USA. There was the concrete concern that commercial banking business and the payments system were suffering losses from equity markets volatility. The rationale of the legislative measure hence was to limit the bank lending to productive uses (in favour of industry, commerce, agriculture, etc.), avoiding speculation in financial markets and on this basis restoring confidence in the soundness of banks.

To this end, it was decided to separate commercial from investment banks. The former, focused on taking deposits and making loans, were prohibited from underwriting or dealing in securities (with the exception of US government securities and other similar securities with no risk). On the other side, investment banks, focused on the underwriting and dealing in securities, were no longer allowed to have connections (e.g., common ownership and/or overlapping directors) with commercial banks.

As a result, commercial and investment banks had to decide to be one or the other, but not both. Some banks decided to remain only commercial (it was the case, among others, of Citibank, Chase Manhattan Bank). Other banks, on the other hand, decided to maintain a position as a commercial bank through its subsidiary (e.g., JP Morgan), while some partners left for the investment Morgan Stanley.

Another important step in the evolutionary path of commercial and investment banking is represented by the Gramm-Leach-Bliley Act of the 1999, which repealed the Glass-Steagall Act. In particular, it decided that commercial banks, investment banks and insurance companies could affiliate with each other becoming part of a financial holding company. The final goal was to support financial integration, started during the previous years, even from a legislative point of view. In those years, most investment banks merged with commercial banks. This was a way for exploiting on one side the investment banks' know-how and expertise about the financial markets and on the other the commercial banks' balance sheet (capital above all) in order to reach an attractive operational mix.

Investment banks, as the most relevant financial players in the direct market, present some important selling points that are the capacity to price a deal and to rely on strong distribution networks. As said, they are focused on supporting firms issuing new debt or equity securities in the primary market aimed at financing their investments. Once, the securities are sold; investment banks may play a role also as brokers or dealers in the secondary market. Particularly, the brokerage function provides that buyers and sellers of securities are brought together and from this the intermediary earns a commission. On the other hand, the dealer function involves a wider role, as the intermediary in this case executes buy and sell orders and trades for its own account according to customers liquidity preferences. For this, the intermediary is also called market maker, and it aims at gaining a positive difference from the selling price and the buying price of securities (Morrison & Wilhelm, 2007).

The issue of securities to the market can be of different types.

It is a primary offering when stocks or bonds are newly issued. In the case it is the first time that a firm offers new securities to the market we have an initial primary offering (IPO), while if the firm has securities already traded in the market we have a seasoned offering (Deloof et al., 2009).

Besides, the securities offering could be a public offering or a private placement (Beatty & Ritter, 1986; Tinic, 1988). In the case of a public offering, firms should choose between an underwritten issue and the best-efforts offering. The former is characterized by the fact that the investment bank is called to ensure the firm the payment of a certain amount of money regardless of whether securities are sold or not. In this way, the investment bank takes on the risk of an unsuccessful issue. On the other hand, the latter does not provide for any guarantee but only that the investment bank will make its best-efforts in selling the securities. Here, the investment bank does not take on the risk and it is generally remunerated according to the number of securities sold. It is worth noting that generally in a public offering a crucial element is the price determination, as securities are traded for the first time. Hence, the selling price of securities is a key-factor for the issuer because the financial resources which can be gathered depend also on it. At the same time, if the price is too high the securities demand could be insufficient and the offering could be cancelled or the underwriter could not be able to sell securities at the agreed price. In this case, the result for the investment bank may be a loss.

Another alternative is between the solicit investment banking services, when the firm publicly announces the intention of selling securities and therefore solicits offers from several intermediaries through competitive bidding and the direct negotiation with a certain investment banks. The latter is generally the mostly used form because of the great deal of work to be accomplished. In particular, there is the need for the investment bank to spend time and resources in performing due diligence analysis about the potential issuers, aimed at gathering and processing all the available data and information before selling securities to the market. The presence of an appropriate due diligence by a specific investment bank is a sort of guarantee for investors who tend to interpret the involvement of the investment bank as a certification value of the issue.

The most important steps of the investment banks activity are the origination, underwriting and distribution. During the origination, the role of the investment bank is to determine whether the operation meets the conditions to be viable. Among the factors to be considered there are the total amount of the financing, its type (debt versus equity), as well as the main features of the securities to be issued (i.e. maturity, rate, etc.). Then we have the underwriting, that is when the investment bank guarantees to buy the securities for the agreed price, taking on the risk for the period ranging from the time it buys the securities from the issuer and the time it sells them to the market. The last step is the selling of the securities by the investment bank to the market (distribution), where investors could be professional/institutional investors (i.e. pension funds, mutual funds, etc.), which generally buy large blocks of securities or retail investors (i.e. individuals and firms), which are mostly interested in small blocks of securities.

Lastly, we have a private placement when the issuer sells the securities directly to the market. Here, the investment bank plays a more limited role. Hence, there is not the underwriting activity but only the investment bank involvement, remunerated by a fee, in bringing buyer and seller together in order to close the transaction at a certain price. Generally, the private placement is reserved to a group of investors.

Basically, private placement is less expensive compared to public offering, as the former implies lower costs and shorter time. There are less disclosure obligations for the issuer, but securities could be less liquid.

4.4 Other Business Areas

Other products and services performed by banks in favour of firms may include, among others, asset management, merger and acquisition (M&A) services and syndicated loans (Allen et al., 2004; Gustafson et al., 2021; Simons, 1993).

Asset management may vary from a less to a more sophisticated offer of services regarding customers' investments. Hence, it may consist of simple investment information or value-added advisory services to customers interested in allocating capital to a particular investment requiring an in-depth analysis in order to ensure the best risk-return mix. Asset management provides a fee to the intermediary (Gründl & Gal, 2017).

Among the services that a bank could offer to firms, there are also the M&A services that are aimed at supporting firms in identifying potential candidates that match the acquiring firm's needs (Allen et al., 2004). This is intuitively easy when the intermediary may boast a robust network and many contacts at regional and international level. The goal is to assess the feasibility of the acquisition by analysing all the features of the acquisition candidate (among others, financial statement, expected cash flows, management, etc.) in order to price the deal. The overall activity linked to the negotiation of the deal is hence the contribution offered to the acquiring firm by the intermediary. Lastly, once the deal is complete; it helps the firm to raise funds to finance the acquisition.

Syndicated loans were developed at the beginning of the late 1970s in the most advanced financial systems on an international scale (Gustafson et al., 2021; Simons, 1993). They provide that some banks (a pool), whose number depends on the extent of the firm's financial needs, are involved in the release of funds. In other words, they are loans granted on the basis of an agreement between some banks and of a single loan agreement. They are hence granted by a pool of banks to a certain firm, generally large-sized.

It is worth noting that they allow banks to diversify their loans portfolio with reference to the industry and the geographic area to which firms belong as well as to reduce the risks linked to large-value transactions. Besides, particularly for smaller intermediaries, they allow the involvement in important and significant transactions, otherwise precluded.

It should be clear that such loans do not have particular technical peculiarities in comparison with other types of banking loans; rather, their specificities concern the ways they are granted and the organization of the transactions. To this end, it should be underlined again the presence of a pool of banks, with a lead bank playing a significant role. Generally, such lead banks are of a big size and have good standing, qualified skills and expertise; for this they are able to deal with the management of the relationship with the corporate borrower. In addition to the agreement among all the banks of the pool on the basis of which all the banks participate to the granting of the loan, there is the loan agreement concluded between the lead bank and the corporate borrower where conditions and terms are defined.

Generally, a syndicated loan starts with the identification of the arranger that is the bank that should create the pool of banks. To this end, it is necessary to prepare an information memorandum where the technical features of the loan (amount, maturity, interest rate, etc.) as well as the general standing of the debtor are reported. This is important for giving all the information to the potential intermediaries interested in the pool involvement.

Banks participating in the pool may have different roles. There is the bank arranger, which manages the organization of the loan. The lead bank is a primary standing intermediary and its participation is attractive for the involvement of others. Then there are the banks which guarantee the underwriting and the participants which guarantee the granting of the loan. The management of the loans is on charge of the agent bank.

References

Allen, L., Jagtiani, J., Peristiani, S., & Saunders, A. (2004). The role of bank advisors in mergers and acquisitions. *Journal of Money, Credit and Banking*, 197–224. https://www.jstor.org/stable/3839017

European Central Bank—ECB (2013). Non-performing loans What matters in addition to the economic cycle? https://www.ecb.europa.eu/pub/pdf/scpwps/ecbwp1515.pdf

Bank for International Settlements—BIS (2013). Financial crises and bank funding: recent experience in the euro area. https://www.bis.org/publ/work406.pdf

European Central Bank—ECB (2017). Guidance to banks on non-performing loans. https://www.bankingsupervision.europa.eu/ecb/pub/pdf/guidance_on_npl.en.pdf

Bank for International Settlements—BIS (2018). Structural changes in banking after the crisis. https://www.bis.org/publ/cgfs60.pdf

European Central Bank—ECB (2020a). Loan types and the bank lending channel. https://www.ecb.europa.eu/pub/pdf/scpwps/ecb.wp2409~79137cac1e.en.pdf

European Central Bank—ECB (2020b). Working Paper Series Bank Lending in the Knowledge Economy. https://www.ecb.europa.eu/pub/pdf/scpwps/ecb.wp2429~c7e30ec0e2.en.pdf

Beatty, R. P., & Ritter, J. R. (1986). Investment banking, reputation, and the underpricing of initial public offerings. *Journal of Financial Economics, 15*(1–2), 213–232. https://doi.org/10.1016/0304-405X(86)90055-3

Cucinelli, D., Di Battista, M. L., Marchese, M., & Nieri, L. (2018). Credit risk in European banks: The bright side of the internal ratings based approach. *Journal of Banking & Finance, 93*, 213–229. https://doi.org/10.1016/j.jbankfin.2018.06.014

Deloof, M., De Maeseneire, W., & Inghelbrecht, K. (2009). How do investment banks value initial public offerings (IPOs)? *Journal of Business Finance & Accounting, 36*(1–2), 130–160. https://doi.org/10.1111/j.1468-5957.2008.02117.x

Doumpos, M., Lemonakis, C., Niklis, D., & Zopounidis, C. (2019). Analytical techniques in the assessment of credit risk. *EURO Advanced Tutorials on Operational Research. Cham: Springer International Publishing.[Google Scholar].*

European Banking Authority—EBA. (2020). Guidelines on loan origination and monitoring. https://www.eba.europa.eu/regulation-and-policy/credit-risk/guidelines-on-loan-origination-and-monitoring

Ferretti, P. (2016). *New perspectives on the bank-firm relationship*. Palgrave Macmillan.

Forestieri, G. (2018). Corporate & investment banking. *Corporate & investment banking*, 1–372.

Gründl, H., & Gal, J. (2017). The evolution of insurer portfolio investment strategies for long-term investing. *OECD Journal: Financial Market Trends, 2016*(2), 1–55. https://doi.org/10.1787/19952872

Gustafson, M. T., Ivanov, I. T., & Meisenzahl, R. R. (2021). Bank monitoring: Evidence from syndicated loans. *Journal of Financial Economics, 139*(2), 452–477. https://doi.org/10.1016/j.jfineco.2020.08.017

Howcroft, B., Durkin, M., Armstrong, G., & Emerson, E. (2007). Small business–bank relationships and the role of Internet banking. *The Service Industries Journal, 27*(7), 947–961. https://doi.org/10.1080/02642060701570784

International Monetary Fund—IMF (2022). The globalization of finance. https://www.imf.org/external/pubs/ft/fandd/2002/03/hausler.htm

Mergaerts, F., & Vander Vennet, R. (2016). Business models and bank performance: A long-term perspective. *Journal of Financial Stability, 22*, 57–75. https://doi.org/10.1016/j.jfs.2015.12.002

Meslier-Crouzille, C., Morgan, D. P., Samolyk, K., & Tarazi, A. (2015). The benefits of geographic diversification in banking. https://core.ac.uk/download/pdf/52862612.pdf

Montedoro (2020). The financial structure of SMEs from 2017 to date. https://www.bgt-grantthornton.it/en/tophic-newsletter/june-2020/the-financial-structure-of-smes-from-2017-to-date/

Morrison, A. D., & Wilhelm, W. J. (2007). Investment of banking: Past, present, and future. *Journal of Applied Corporate Finance, 19*(1), 42. https://international.hunters.com/download/a17e9_investment-banking-past-present-and-future

Naili, M., & Lahrichi, Y. (2022). Banks' credit risk, systematic determinants and specific factors: Recent evidence from emerging markets. *Heliyon, 8*(2), e08960. https://doi.org/10.1016/j.heliyon.2022.e08960

OECD (2013). The role of banks, equity markets and institutional investors In long-term financing for growth and development. https://www.oecd.org/finance/private-pensions/G20reportLTFinancingForGrowthRussianPresidency2013.pdf

OECD (2018). Enhancing SME access to diversified financing instruments. https://www.oecd.org/cfe/smes/ministerial/documents/2018-SME-Ministerial-Conference-Plenary-Session-2.pdf

OECD (2019). Due diligence for responsible corporate lending and securities underwriting: Key considerations for banks implementing the OECD Guidelines for Multinational Enterprises. https://mneguidelines.oecd.org/final-master-due-diligence-for-responsible-corporate-lending-and-securities-underwriting.pdf

Shim, J. (2019). Loan portfolio diversification, market structure and bank stability. *Journal of Banking & Finance, 104*, 103–115. https://doi.org/10.1016/j.jbankfin.2019.04.006

Simons, K. (1993). Why do banks syndicate loans?. *New England Economic Review*, (Jan), 45–52. https://ideas.repec.org/a/fip/fedbne/y1993ijanp45-52.html

Song, H., Yu, K., & Lu, Q. (2018). Financial service providers and banks' role in helping SMEs to access finance. *International Journal of Physical Distribution & Logistics Management, 48*(1), 69–92. https://doi.org/10.1108/IJPDLM-11-2016-0315

Tinic, S. M. (1988). Anatomy of initial public offerings of common stock. *The Journal of Finance, 43*(4), 789–822. https://doi.org/10.1111/j.1540-6261.1988.tb02606.x

Van Deventer, D. R., Imai, K., & Mesler, M. (2013). *Advanced financial risk management: Tools and techniques for integrated credit risk and interest rate risk management*. John Wiley & Sons.

Vo, X. V. (2018). Bank lending behavior in emerging markets. *Finance Research Letters, 27*, 129–134. https://doi.org/10.1016/j.frl.2018.02.011

CHAPTER 5

FinTech: Challenges and Opportunities for Banks and Financial Markets

Abstract This chapter introduces the phenomena of FinTech, providing a discussion about its main implications for banks and financial markets. Firstly, this chapter examines the implications for the banking industry, highlighting the opportunities as well as the risks that FinTech poses to banks' business. Secondly, it explores new sources of financing focusing on the most important innovations in the market for entrepreneurial finance, namely crowdfunding and Initial Coin Offerings (ICOs).

Keywords FinTech · Banks · Entrepreneurial finance · Crowdfunding · Initial coin offerings

5.1 The Rise of FinTech

Recent technological developments have given rise to the FinTech industry, which covers digital innovations and technology-enabled business model innovations in the financial sector (Allen et al., 2021; Frame et al., 2018; Goldstein et al., 2019; Philippon, 2016). Fundamentally, FinTech concerns the use of technology to provide new and enhanced financial services (Thakor, 2020). Some examples of such innovations and new technologies included in the FinTech space are distributed ledger technologies (DLTs) (e.g., blockchain), digital currencies (e.g.,

P. Ferretti and P. Martino, *Banking and Financial Markets*,
https://doi.org/10.1007/978-3-031-32562-5_5

cryptocurrencies), new digital advisory and trading systems, artificial intelligence (AI) and machine learning, peer-to-peer lending and equity crowdfunding. These innovations hold potential benefits for all parties involved in financial services: they may expand access to financial services and reach under-served consumers (i.e. *financial inclusion*); they may reduce transaction costs and provide greater transparency, in addition to providing greater convenience and efficiency. This is also acknowledged by the European Banking Authority (EBA), which stresses the potential of these new technological innovations, pointing out that they may bring many benefits to consumers and organizations, including access to credit, improved comparability of products, access to a wider product range, availability of up-to-date information, tailored product offerings, reduced costs and consumer convenience (EBA, 2017).

The FinTech industry has evolved significantly over the last decade, experiencing a massive year of investment globally in 2021, with total global investments (across venture capital [VC], private equity [PE] and mergers and acquisitions [M&A]) amounting to $226.5 billion (KPMG, 2022). As shown in Fig. 5.1, global investments in FinTech in 2022 (as of 30 June 2022) fell to $107.8 billion across 2980 deals due to several factors (i.e. the uncertainties related to the Russia-Ukraine conflict, ongoing supply chain challenges, as well as rising inflation and interest rates), although remaining to levels seen in previous years thus highlighting the strength of the global FinTech market.

So, what is so special about the current FinTech revolution? Of course, technology has always influenced the financial industry, with advancements changing the way the financial industry operates (Berger, 2003; Frame & White, 2014). Nevertheless, today we are entering a new digital age with new technologies having the potential to change the financial industry in faster and more significant ways than ever before. In order to understand the potential of FinTech, it is useful to recall the working definition by the Financial Stability Board (FSB)'s (2017), which indicates FinTech as "technologically enabled financial innovation that could result in new business models, applications, processes or products with an associated material effect on financial markets and institutions and the provision of financial services". What emerges from this definition is the *disruptive nature* of these new innovations or, in other words, their ability to have a material effect on the structure and functioning of the financial system (both the direct and indirect financing channels), posing new risks as well as new opportunities. In particular, according to Philippon (2016)

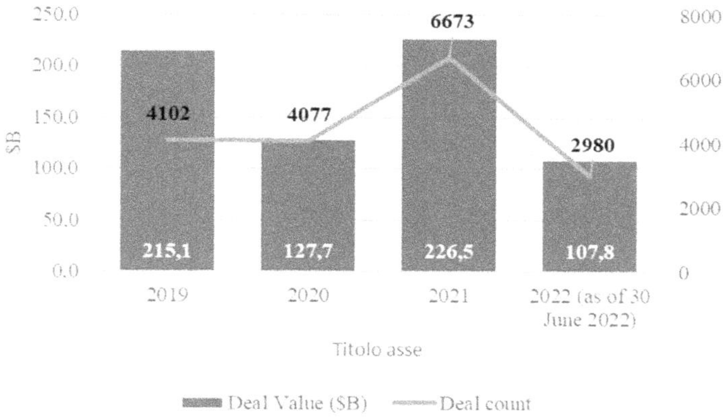

Fig. 5.1 Total global investment activity (VC, PE, and M&A) in FinTech (*Source* Own processing of data from KPMG [2022], *Pulse of FinTech H1 2022*)

these innovations and new technologies have the potential to: (1) facilitate the strategic disintermediation; (2) revolutionize how existing firms create and deliver products and services; (3) democratize access to financial services, especially in providing new gateways for entrepreneurship. This means that FinTech may contribute to the increase in efficiency and competition in the financial industry, as well as provide wider access to financial services by offering new ways to obtain funds.

In the next paragraphs, the aforementioned implications of FinTech will be outlined with specific regard to banks and financial markets, especially the entrepreneurial finance market.

5.2 FinTech: Implications for Banks

The rapid development of FinTech has brought opportunities and challenges to banks (Boot et al., 2021; Zhao et al., 2022). On the one hand, new technologies and innovations are transforming bank products and services, as well as bank processes, resulting in a more efficient banking system with new payment tools and bank services emerging (Frame & White, 2014; Frame et al., 2018). On the other hand, new actors—referred to as "fintech firms"—have recently emerged and offer traditional banking products and services in an innovative way, thereby posing new threats to banks.

5.2.1 FinTech and Banking Competition

One of the main implications of FinTech is that it can facilitate the strategic disintermediation (Philippon, 2016). *Disintermediation* means eliminating the need for intermediaries such as banks in transactions or, in other words, bypassing traditional intermediaries in the offering of financial services. This is because new players referred to as FinTech firms have recently emerged and offer products and services that traditionally were banking products, thereby increasing the competition in the industry (Berg et al., 2022; Cornelli et al., 2023; EBA, 2018; Martino, 2021; Stulz, 2019). Notably, thanks to the use of new technologies (e.g., blockchain, cryptocurrencies, digital platforms, etc.) these new firms are able to offer banking products and services in innovative ways, offering better user experience, and at a lower price thus causing traditional financial intermediaries to potentially lose their market shares. The *P2P lending*—also referred to as *marketplace lending*—is an example of innovation in this area, representing the provision of loans to individuals and businesses through online services that directly match lenders with borrowers without using a bank.

Hence, the considerations above suggest viewing FinTech as a threat for banks, since new competitors are entering the market which may jeopardize some of the core businesses of banks, i.e. lending and payments, with the consequent erosion of bank revenues, thus potentially undermining their profitability (Martino, 2021; Yeoh, 2017). Although the involvement of non-financial firms in the provision of financial services is not novel per se, the rise of FinTech firms may change this fundamentally, since they are able to offer a vast range of products and services, and at the same time appear to have an edge over banks which have bigger, less flexible and more expensive infrastructures (Boot et al., 2021). According to the EBA (2017), while alternative lending platforms can put pressure on the interest income from loans of existing credit institutions, new entrants offering commoditized products and services (such as money transfers and brokerage), at lower costs, can reduce the fees and commissions income of established players. In this respect, several studies show that many FinTech firms are active today in segments of the value chain already managed by banks, such as business and consumer lending (Cornelli et al., 2023; Gopal & Schnabl, 2022; PWC, 2020). Such firms leverage new technologies to offer credits and funding to individual and business, thus competing with traditional commercial banks. Moreover, they are increasingly reshaping the way payments are made, changing how consumers and companies purchase goods and services and make payments (Ehrentraud et al., 2021). The most notable example in this area are cryptocurrencies like Bitcoin—the first and most-known cryptocurrency (Böhme et al., 2015; Giudici et al., 2020; Yermack, 2015)—which allows online payments to be sent directly from one party to another, without going through a financial institution (Nakamoto, 2008). Today, the number of payments made using cryptocurrencies is consistently on the rise, with several companies (such as Microsoft, Expedia, etc.) accepting Bitcoin and other cryptocurrencies, and others allowing customers make cryptocurrency payments via crypto payment gateways such as Bitpay—a cryptocurrency payment processor that enables to accept cryptocurrencies as a payment method.

Finally, it is worth noting that FinTech may also hit banks on the costs side (Martino, 2021). Indeed, these new competitors can create pressure on banks to modify or adapt existing products and services in order to stay competitive and affect the willingness of customers to do business with banks, requiring substantial expenditures (investments) that put a burden in the income statement of banks.

To date, the size of the competition is still limited, but activity is accelerating as shown in the latest report by the Cambridge Centre for Alternative Finance (CCAF). In this respect, Fig. 5.2 shows that, globally, transactions via FinTech platforms have experienced significant growth over the years, with platform activity reaching its peak at USD 419 billion in 2017. As shown in Fig. 5.2 market development followed very different paths in China and the rest of the world, with the former that experienced a quick and dramatic cycle of boom and bust. Notably, while in the earlier years the role of the Chinese P2P lending industry served as a substantive and dominant driver of total transaction volume, making up the largest market shares and growing at a considerable pace, the prominence of the Chinese lending marketplace has subsequently decreased significantly due to regulatory changes introduced in 2018. By contrast, when observing the rest of the world total transaction volumes attributed to alternative finance platforms followed a steady and gradual growth trajectory, with the global volumes that rose from $44 billion in 2015 to $113 billion in 2020.

Looking at the geographic distribution in 2020, the United States and Canada amount to 65% of global market volumes, followed by the United

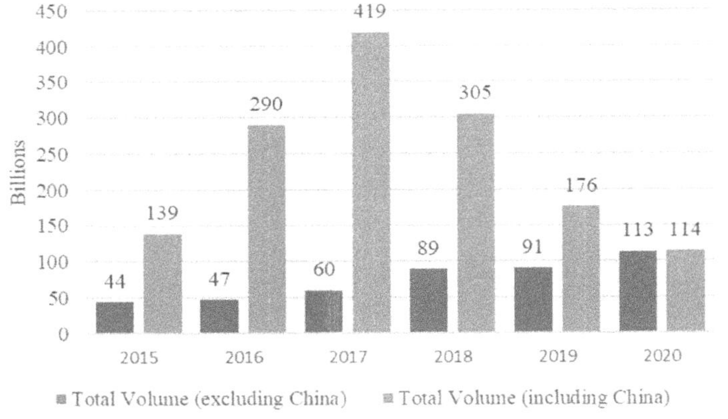

Fig. 5.2 Total global alternative finance volume 2015–2020, USD (*Source* Own processing of data from "The 2nd Global Alternative Finance Market Benchmarking Report" by the Cambridge Centre for Alternative Finance [CCAF])

Kingdom (11%) and Europe (9%), as reported in Fig. 5.3. With specific regard to the European context, Fig. 5.4 shows that market volumes (including the UK) grew consistently from $1.5 billion in 2013 to $22.6 billion in 2020.

Interestingly, when looking at users of FinTech platforms, Fig. 5.5 shows that it is predominantly ***banked*** individuals and businesses (i.e. users that have access to a full suite of financial services) that are currently using these new sources of financing (i.e. digital lending and capital raising).[1] This demonstrates the potential of FinTech to reach traditional banks' customers, in addition to ***underbanked*** (users with access to some basic financial services, but not a complete suite) or ***unbanked*** (users not served by or with access to any traditional financial service) users, meaning that it may threaten banks' business. A recent study by Eça et al. (2022) concurs, finding that FinTech firms serve high quality and creditworthy small businesses who already have access to bank credit, thereby suggesting that

[1] Data from the CCAF Report also shows the trend towards more intensive use of FinTech platforms by unbanked or underbanked in less developed geographical areas (Sub-Saharan Africa, Asia-Pacific). This highlights the potential of FinTech to facilitate financial inclusion.

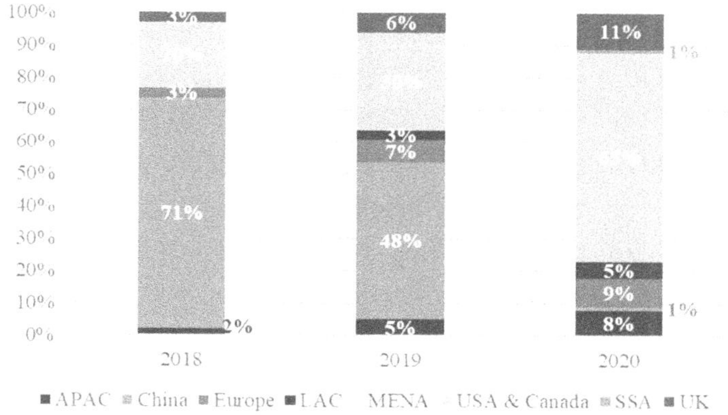

Fig. 5.3 Market share of alternative finance activity by region (*Source* Own processing of data from "The 2nd Global Alternative Finance Market Benchmarking Report" by the Cambridge Centre for Alternative Finance [CCAF])

Fig. 5.4 European online alternative finance market volumes 2013–2020, USD (*Source* Own processing of data from "The 2nd Global Alternative Finance Market Benchmarking Report" by the Cambridge Centre for Alternative Finance [CCAF])

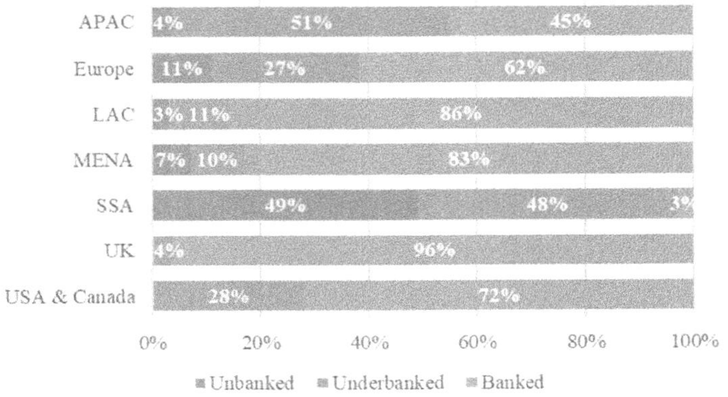

Fig. 5.5 Banking status, by region, of users of FinTech platforms (*Source* Own processing of data from "The 2nd Global Alternative Finance Market Benchmarking Report" by the Cambridge Centre for Alternative Finance [CCAF])

these new players directly compete with banks for the same segment of firms and do not cater to different, riskier firms.

Overall, the reported data document the recent growth of FinTech credit. However, it is worth mentioning that, to date, it is difficult to estimate how large FinTech credit will become in the future and the related implications for banks or, in other words, whether these alternative forms of credit will complement more traditional intermediaries or rather substitute for them (Cornelli et al., 2023; Thakor, 2020).

5.2.2 FinTech and Banking Efficiency

Notwithstanding the potential risks identified above, the literature suggests that FinTech can also bring about certain opportunities for banks to improve their efficiency in terms of costs reduction and/or an increase in revenues (Colombini, 2018), and consequently, to potentially improve their profitability. Notably, scholars argue that the use of new technologies may offer banks solutions to increase cost efficiencies and address users' complex needs, thereby improving their competitive advantage (Lee et al., 2021; Martino, 2019; Yermack & Fingerhut, 2019). For example, Lee et al. (2021) identify multiple channels through which

FinTech can affect bank efficiency. They suggest that FinTech can facilitate financial innovation and shape the way banks conduct business, thereby affecting bank efficiency as the application of new technologies is expected to reduce bank costs over time (e.g., through the automatization of internal processes); it can also change the delivery channels for financial services, as well as promote the development of new and more complex financial products. The European Supervisors Authorities (ESAs) also acknowledged the potential of FinTech for banks, suggesting that the use of new technologies can allow financial intermediaries to develop and distribute their products and services to a wider consumer base, potentially faster and at a lower cost, in turn reducing barriers to entry and time to market for new products, and fostering financial innovation (ESA, 2022). Moreover, by leveraging on the wide range of consumer data made available through digitalization and advanced data analytics techniques, such as big data analytics, machine learning and AI,[2] financial intermediaries may be capable of better understanding consumers' needs and preferences and customizing their marketing approaches and products accordingly, thereby boosting the personalization of financial products and services.

Hence, if banks are able to integrate new technologies into their business model in order to provide their services, they may exploit their benefits to improve banks efficiency in terms of cost and time reduction of operations.[3] This can also be instrumental to improving the quality of the services offered to clients, thereby providing an important competitive edge. The considerations above are particularly interesting in terms of banks' potential profitability as new technologies may lower banks' operational costs, by improving banking processes' efficiency, and increasing bank revenues through the provision of new banking products and services.

[2] Artificial intelligence refers to the general ability of computers to emulate human thought and perform tasks in real-world environments, while machine learning refers to the technologies and algorithms that enable systems to identify patterns, make decisions and improve themselves through experience and data. Machine learning is considered a subset of AI. Big data analytics, on the other hand, refer to the methods, tools and applications used to collect, process and derive insights from varied, high-volume, high-velocity data sets.

[3] For a general overview about potential use cases of these new technologies, see the EBA (2018) *Report on the prudential risks and opportunities arising for institutions from FinTech*.

To date, many EU banks are embarking on digitalization projects aimed at streamlining and automating their back office operations in various areas ranging from trading in financial products, accounting and loan processing and administration to automated analysis and decision-making and compliance (Boot et al., 2021; EBA 2017; Martino, 2021). Some banks are setting up innovation and accelerator hubs and entering into direct competition with FinTech firms developing new, digital-oriented financial products and services that match customer needs (e.g., online trading platforms and alternative lending arrangements like P2P lending), with the aim of either protecting their existing revenues from the FinTech competition or achieving new revenues to increase their fees and commissions income. Other banks are acquiring or collaborating with FinTech firms and entering into partnerships with start-ups and technology firms in order to procure expertise for the digitalization of key processes, as well as to offer new and digital products to their existing customers (see Box 5.1).

Box 5.1: An Example of Collaboration Between Banks and FinTech Firms

Among the several initiatives undertaken to face the FinTech challenge, the Italian banking group Intesa Sanpaolo invested in 2022 £40 million in Thought Machine (as its technology partner), the next-generation cloud-based banktech company, to create Isybank—the Group's new digital banking platform—whose aim is to initially serve—with a wide range of digital banking services—the 4 million Intesa Sanpaolo retail customers in Italy who do not use the branch network.

The growing attention of banks in FinTech technologies is also confirmed by the latest report on Risk assessments of the European Banking System by the EBA (2022), which shows the level of involvement of EU banks in several FinTech's technologies. As shown in Fig. 5.6, overall, the level of adoption of new technologies by banks has increased over the years. As an example, the use of artificial intelligence (AI) solutions (including machine learning and natural language processing [NLP]) continues to increase, with 83% of banks responding to the Risk Assessment Questionnaire (RAQ) reported that they already use AI (including machine learning and NLP), and an additional 12% is either pilot testing or developing AI systems for various use cases. Among them, the most common use cases of AI/machine learning are (1) fraud detection (82%), (2) Anti-Money Laundering/Countering the Financing of Terrorism (AML/CFT) purposes (80%), (3) creditworthiness assessment or credit scoring (80%) or (4) profiling/clustering of clients or transactions (77%) (EBA, 2022). Other popular AI applications relate to real-time monitoring of payments, risk modelling, including regulatory credit risk modelling or conduct risk monitoring. The use of cloud computing has also increased, with 85% of RAQ respondents reported it to be in use, up from 71% in 2021, while changes in the use of other monitored financial technologies has been less pronounced. Interestingly, the new data collected on the use of application programming interfaces (APIs) and quantum computing indicate that almost all banks (95%) are already using APIs, while the use of quantum computing is at a very early stage—3% of banks reported it in use, additional 7% in pilot testing.

The considerations above, taken together, provide a general overview on the main implications of FinTech for banks in terms of opportunities and risks. To conclude, it is worth mentioning the key role that *regulation* plays in the development of FinTech and its impact on the banking industry. On the one hand, regulations (e.g., capital, liquidity and risk-management requirements) for FinTech firms are necessary to foster innovation and/or competition while not compromising other policy objectives such as financial stability and customer protection (Ehrentraud

Fig. 5.6 Level of involvement of banks with the application of the selected technologies (*Source* Own processing of data from the EBA's [2022] risk assessment report of the European banking system)

et al., 2021); on the other, the revision of existing rules and/or the introduction of new ones is crucial to promote the development and implementation of new technologies by banks in compliance with the existing regulatory framework (e.g., personal data protection). Hence, a clear and balanced regulatory framework is needed to enable FinTech to flourish and, at the same time, ensuring the safety and soundness of the banking system (and more in general of the financial industry), including consumer and other stakeholders protections.

5.3 FinTech: Implications for the Entrepreneurial Finance Market

Financial markets are affected by several problems (i.e. information asymmetry, uncertainty and transaction costs) which make the selection process, matching and monitoring difficult, preventing access to finance by some firms, such as entrepreneurial ventures. FinTech can affect the functioning of financial markets in different ways, providing new innovations which potentially improve access to finance for firms and afford new possibilities for investors (Bollaert et al., 2021). This is particularly true for the market for entrepreneurial finance, which includes a wide array of sources of capital, such as venture capital (VC), angel investors, equity and debt finance. Notably, the emergence of new innovation technologies has prompted the development of new financial alternatives for seeding entrepreneurship, with many new players and new ways to finance entrepreneurial firms that have entered the arena (Bellavitis et al., 2017; Bertoni et al., 2022; Block et al., 2021; Martino et al., 2022). The most prominent examples of such innovative forms of entrepreneurial finance are crowdfunding and Initial Coin Offerings (ICOs), which have emerged as two distinct but important entrepreneurial finance segments of today's ecosystem (Block et al., 2021; Bruton et al., 2015).

5.3.1 Crowdfunding

According to Belleflamme et al. (2010, p. 5), crowdfunding involves "an open call, essentially through the Internet, for the provision of financial resources either in form of donation or in exchange for some form of reward and/or voting rights". By leveraging the geographic and social reach of the internet to connect fundraisers to millions of potential backers (i.e. investors), crowdfunding allows entrepreneurs to obtain funds from a large audience (i.e. the crowd), where each individual will provide a very small amount instead of raising the money from a small group of sophisticated investors (Belleflamme et al., 2014; Bruton et al., 2015; Fleming & Sorenson, 2016; Moritz & Block, 2016).

In a crowdfunding campaign, the process of fundraising takes place on the so-called crowdfunding platforms, i.e. internet-based platforms which act as intermediaries between the fundraiser (individuals, start-ups or companies) and potential investors. Specifically, such platforms allow entrepreneurs to advertise and pitch their products and ideas to the community of online investors, in addition to providing infrastructures for managing payments and for keeping track of and communicating with backers (Agrawal et al., 2011; Fleming & Sorenson, 2016). In exchange for the services provided, platforms usually charge those receiving funds a fee, typically a percentage of the amount raised.

The literature identifies four main models of crowdfunding, namely: *reward*, *donation*, *lending* and *equity* crowdfunding (e.g., Belleflamme et al., 2014; Block et al., 2018; Martino et al., 2022). In *reward*-based crowdfunding, entrepreneurs raise funds offering backers a reward for their support, which typically involves the delivery of a product or service of the company that will be commercialized if the project is successful. The *donation*-based crowdfunding entails no remuneration in exchange of the money pledged by backers; here, proponents are typically individuals or non-governmental organizations raising money to support humanitarian and artistic projects, while funders act as philanthropists who donate for charitable giving and/or social image. Contrary to the two abovementioned models, where funders do not receive a monetary compensation, in *equity* and *lending*-based crowdfunding backers are investors in a campaign who may obtain monetary benefits. Specifically, in *lending*-based crowdfunding the fundraiser borrows capital from the crowd in the form of loans, while funders receive interests in exchange for their capital provided. Finally, in the *equity*-based crowdfunding the fundraiser offers equity stakes in exchange for the capital invested by the crowd. As such, the proponent is by definition a company (Vismara, 2018), while funders are shareholders who acquire ownership and voting rights with the intent to participate in the distribution of future profits (Ahlers et al., 2015). The equity-based crowdfunding is in most jurisdictions subject to securities laws.

5.3.2 *Initial Coin Offerings (ICOs)*

Similarly to crowdfunding, ICOs enable new ventures to raise money from the public. Specifically, an ICO can be defined as an event where a start-up or project sells its new cryptocurrency for the first time to the public in order to raise capital (Adhami et al., 2018; Fisch, 2019). In an ICO, the demand for capital usually consists of digital entrepreneurs aiming to raise capital to launch a new business or to foster the growth of their entrepreneurial ventures, especially in the blockchain industry (Fisch, 2019). However, it has the potential to reach traditional business as well.

To start an ICO, the start-up or project team will usually publish a document called a "*white paper*", which describes the details of the project, including, for example: the cryptocurrencies that they are going to offer, a description of the project (e.g., team description, business plan, development roadmap), as well as the technical specifications and ICO details (Adhami et al., 2018).

Although an ICO presents similarities with crowdfunding, it has some key features that make it different from crowdfunding and other forms of entrepreneurial finance: in particular, its offering has a different nature—i.e. tokens—and it uses blockchain technology for verification instead of crowdfunding platforms (Martino et al., 2020a, 2020b).

Firstly, a key feature of ICOs is that they sell *tokens* to investors, which represent blockchain-based digital assets. Specifically, a token corresponds to a unit of value issued by a venture and covers a wide range of applications (Bellavitis et al., 2021; Fisch, 2019; Howell et al., 2020): it can be an *utility token* which assigns a right to investors to redeem the token for a company's product or service once developed, or a *security token*, which resembles traditional financial investments and has an underlying investment asset that investors acquire. As equity crowdfunding, the offering of security tokens—called Security Token Offerings (STOs)—is in most jurisdictions subject to securities laws. Beyond the distinction between utility and security tokens, it is worth noting that often all tokens are cryptocurrencies that are meant to function as a currency in the venture's own ecosystem (Fisch, 2019).

Another key feature of ICOs is that they work on a *blockchain technology*—the distributed ledger technology behind bitcoin (and other cryptocurrencies)—on which tokens are issued and sold. The blockchain technology can be defined as a digital, decentralized and distributed ledger that is constantly updated (Martino, 2021; Yermack, 2017). It represents a central innovation in the FinTech space, enabling a novel approach to recording and transmitting data across a network in an immutable manner. The blockchain technology serves as a processing platform, enabling a direct transaction between ICO investors and ICO firms via peer-to-peer financing (Adhami et al., 2018; Bellavitis et al., 2021; Howell et al., 2020; Tapscott & Tapscott, 2017). More specifically, the distributed ledger technology of blockchain enables entrepreneurs to raise funding without involving any intermediaries, since investors can buy tokens directly from the ICO-conducting venture, resulting in a complete disintermediation of the financing process[4] (Momtaz, 2021).

It is widely acknowledged in the literature that both crowdfunding and ICOs have the potential to expand access to financial resources to those often excluded from traditional forms of entrepreneurial finance or, in other words, to underrepresented groups of potential entrepreneurs such as remotely located companies, female entrepreneurs, minority entrepreneurs (in terms of ethnicity), as well as companies with younger top management team (TMT) members (Cumming et al., 2021; Fisch et al., 2022). Compared to traditional forms of entrepreneurial finance, both crowdfunding and ICOs dramatically lower the cost of fundraising[5] (Agrawal et al., 2015) and, by leveraging the geographic and social reach of the Internet, they can overcome the distance-related economic frictions usually associated with financing entrepreneurial ventures, thereby helping overcome an early-stage gap of start-ups (Agrawal et al., 2011; Fisch et al., 2022; Fleming & Sorenson, 2016; Howell et al., 2020; Martino et al., 2022). In addition, both crowdfunding and ICOs may provide benefits for investors as well, by expanding the range of investment opportunities. Notably, they allow anyone (e.g., small investors) to invest any amount in

[4] In this respect, it is worth noting that a new variant of ICOs has recently emerged, namely initial exchange offerings (IEOs), which introduces an intermediary platform in the token offerings. Specifically, IEOs rely on cryptocurrency exchanges to ensure the trustworthiness of potential projects and to connect high-quality projects to potential investors (see, for example, Chen & Bellavitis, 2020). However, to date, IEOs represent a tiny portion of the overall market.

[5] Regarding the cost of fundraising, it is worth mentioning that ICOs allow to further reduce costs included in fundraising by avoiding compliance and intermediary costs thanks to the blockchain technology, which eliminates the middleman such as crowdfunding platforms.

a company, thus dramatically reducing entry barriers to participate financially in the successes of start-up companies (i.e. democratize participation in financial markets).

Notwithstanding the abovementioned benefits for entrepreneurs and investors, it is also worth mentioning potential risks that both crowdfunding and ICOs can pose, particularly for investors. In contrast to other sources of entrepreneurial finance, both crowdfunding and ICOs are currently characterized by a comparably low degree of regulation which exacerbates uncertainty and potentially increases investment risk (Bellavitis et al., 2020; Cumming et al., 2019; Martino et al., 2022; Rossi et al., 2021). In particular, as Block et al. (2021) note, the regulation of crowdfunding and ICOs is largely contingent on the offering characteristics: while equity/lending crowdfunding and STOs are subject to securities laws in some jurisdictions, the reward/donation-based crowdfunding and utility token offerings are conducted in a legal grey zone with no need to comply with any registration or disclosure requirements. In the EU context, for example, it is useful to recall the Regulation on European Crowdfunding Service Providers (ECSP) for business ([EU] 2020/1503), which lays down uniform rules across the EU for the provision of investment-based and lending-based crowdfunding services related to business financing. These new rules aim to provide an aligned and enhanced investor protection framework, based on clear rules on information disclosures for project owners and crowdfunding platforms, rules on governance and risk management for crowdfunding platforms, as well as strong and harmonized supervisory powers for national authorities overseeing the functioning of crowdfunding platforms. Therefore, depending on how they are structured, crowdfunding and ICOs may not be captured by the existing rules and may fall outside of the regulated space with the level of investor protection being at a minimum as there is only a limited basis (only broader laws) to pursue legal action after the offering. Accordingly, this low degree of regulation makes ICOs and crowdfunding vulnerable to fraud and illicit activities[6] (ESMA, 2017; SEC, 2021). This is particularly true for the ICO industry, where empirical studies show the high prevalence of frauds, such as "exit scams" in which the venture team disappears after raising funds thereby swindling investors, as well as

[6] For more details about risks involved in crowdfunding and ICOs see, for example: the *ESMA alerts investors to the high risks of Initial Coin Offerings (ICOs)* published in 2017, as well as the SEC (2021) *Crowdfunding for investors—Updated investor bulletin*.

the high exposure to market manipulation, such as "pump-and-dump" schemes in which actors coordinate to bid up the price of coins before selling at a profit (Bellavitis et al., 2021; Fisch, 2019; Hornuf et al., 2022; Martino et al., 2022). For instance, a study by Hornuf et al. (2021) investigates the extent of fraud in ICOs identifying, in total, 274 fraud cases within the 1393 ICOs studied: 188 were suspected and 175 were confirmed fraud cases.

Hence, the considerations above underline once again the importance of regulation for the development of FinTech in general, and more specifically for crowdfunding and ICOs, which should aim at reducing the risk of fraud and ensuring better investor protection, without hampering the beneficial effects of these new innovations.

References

Adhami, S., Giudici, G., & Martinazzi, S. (2018). Why do businesses go crypto? An empirical analysis of initial coin offerings. *Journal of Economics and Business, 100*, 64–75.

Agrawal, A. K., Catalini, C., & Goldfarb, A. (2011). *The geography of crowdfunding* (No. w16820). National Bureau of Economic Research.

Agrawal, A., Catalini, C., & Goldfarb, A. (2015). Crowdfunding: Geography, social networks, and the timing of investment decisions. *Journal of Economics & Management Strategy, 24*(2), 253–274.

Ahlers, G. K., Cumming, D., Günther, C., & Schweizer, D. (2015). Signaling in equity crowdfunding. *Entrepreneurship Theory and Practice, 39*(4), 955–980.

Allen, F., Gu, X., & Jagtiani, J. (2021). A survey of fintech research and policy discussion. *Review of Corporate Finance, 1*, 259–339.

Bellavitis, C., Cumming, D., & Vanacker, T. (2020). Ban, boom, and echo! Entrepreneurship and initial coin offerings. *Entrepreneurship Theory and Practice, 46*(5), 1136–1169. https://doi.org/10.1177/1042258720940114

Bellavitis, C., Filatotchev, I., Kamuriwo, D. S., & Vanacker, T. (2017). Entrepreneurial finance: New frontiers of research and practice: Editorial for the special issue Embracing entrepreneurial funding innovations. *Venture Capital, 19*(1–2), 1–16.

Bellavitis, C., Fisch, C., & Wiklund, J. (2021). A comprehensive review of the global development of initial coin offerings (ICOs) and their regulation. *Journal of Business Venturing Insights, 15*, e00213.

Belleflamme, P., Lambert, T., & Schwienbacher, A. (2010, June). Crowdfunding: An industrial organization perspective. In *Prepared for the workshop Digital business models: Understanding strategies'*, held in Paris on June (pp. 25–26).

Belleflamme, P., Lambert, T., & Schwienbacher, A. (2014). Crowdfunding: Tapping the right crowd. *Journal of Business Venturing, 29*(5), 585–609.

Berg, T., Fuster, A., & Puri, M. (2022). Fintech lending. *Annual Review of Financial Economics, 14*, 187–207.

Berger, A. N. (2003). The economic effects of technological progress: Evidence from the banking industry. *Journal of Money, credit and Banking, 35*, 141–176.

Bertoni, F., Bonini, S., Capizzi, V., Colombo, M. G., & Manigart, S. (2022). Digitization in the market for entrepreneurial finance: Innovative business models and new financing channels. *Entrepreneurship Theory and Practice, 46*(5), 1120–1135.

Block, J. H., Colombo, M. G., Cumming, D. J., & Vismara, S. (2018). New players in entrepreneurial finance and why they are there. *Small Business Economics, 50*(2), 239–250.

Block, J. H., Groh, A., Hornuf, L., Vanacker, T., & Vismara, S. (2021). The entrepreneurial finance markets of the future: A comparison of crowdfunding and initial coin offerings. *Small Business Economics, 57*(2), 865–882.

Böhme, R., Christin, N., Edelman, B., & Moore, T. (2015). Bitcoin: Economics, technology, and governance. *Journal of Economic Perspectives, 29*(2), 213–238.

Bollaert, H., Lopez-de-Silanes, F., & Schwienbacher, A. (2021). Fintech and access to finance. *Journal of Corporate Finance, 68*, 101941.

Boot, A., Hoffmann, P., Laeven, L., & Ratnovski, L. (2021). Fintech: what's old, what's new? *Journal of Financial Stability, 53*, 100836.

Bruton, G., Khavul, S., Siegel, D., & Wright, M. (2015). New financial alternatives in seeding entrepreneurship: Microfinance, crowdfunding, and peer–to–peer innovations. *Entrepreneurship Theory and Practice, 39*(1), 9–26.

Cambridge Centre for Alternative Finance. (2021). *The 2nd Global Alternative Finance Market Benchmarking Report*. https://www.jbs.cam.ac.uk/wp-content/uploads/2021/06/ccaf-2021-06-report-2nd-global-alternative-finance-benchmarking-study-report.pdf

Chen, Y., & Bellavitis, C. (2020). Blockchain disruption and decentralized finance: The rise of decentralized business models. *Journal of Business Venturing Insights, 13*, e00151.

Colombini, F. (2018). Banking efficiency in Europe. In F. Colombini (Eds.), *Raising capital or improving risk management and efficiency?* Palgrave Macmillan. https://doi.org/10.1007/978-3-319-71749-4_4

Cornelli, G., Frost, J., Gambacorta, L., Rau, P. R., Wardrop, R., & Ziegler, T. (2023). Fintech and big tech credit: Drivers of the growth of digital lending. *Journal of Banking & Finance, 148*, 106742.

Cumming, D. J., Johan, S., & Pant, A. (2019). Regulation of the crypto-economy: Managing risks, challenges, and regulatory uncertainty. *Journal of Risk and Financial Management, 12*(3), 126.

Cumming, D., Meoli, M., & Vismara, S. (2021). Does equity crowdfunding democratize entrepreneurial finance? *Small Business Economics, 56*(2), 533–552.

EBA. (2017). *Discussion paper on the EBA's approach to financial technology (FinTech).* https://eba.europa.eu/sites/default/documents/files/documents/10180/1919160/7a1b9cda-10ad-4315-91ce-d798230ebd84/EBA%20Discussion%20Paper%20on%20Fintech%20%28EBA-DP-2017-02%29.pdf

EBA. (2018). *Report on the prudential risks and opportunities arising for institutions from FinTech.* https://www.eba.europa.eu/sites/default/documents/files/documents/10180/2270909/02c7859f-576e-421e-b243-a145c0eaa131/Report%20on%20prudential%20risks%20and%20opportunities%20arising%20for%20institutions%20from%20FinTech.pdf

EBA. (2022). *Risk assessments of the European Banking System.* https://www.eba.europa.eu/sites/default/documents/files/document_library/Risk%20Analysis%20and%20Data/Risk%20Assessment%20Reports/2022/RAR/1045298/Risk%20Assessment%20Report%20December%202022.pdf

Eça, A., et al. (2022, February). *The real effects of FinTech lending on SMEs: Evidence from loan applications* (ECB Working Paper Series, No. 2639). https://www.ecb.europa.eu/pub/pdf/scpwps/ecb.wp2639~849f8a2608.en.pdf

Ehrentraud, J., et al. (2021, July). *FinTech and payments: Regulating digital payment services and e-money* (FSI Insights, No. 33). https://www.bis.org/fsi/publ/insights33.pdf

ESA. (2022). Joint European Supervisory Authority response to the European Commission's February 2021 Call for Advice on digital finance and related issues: Regulation and supervision of more fragmented or non-integrated value chains, platforms and bundling of various financial services, and risks of groups combining different activities. https://www.eba.europa.eu/sites/default/documents/files/document_library/Publications/Reports/2022/1026595/ESA%202022%2001%20ESA%20Final%20Report%20on%20Digital%20Finance.pdf

European and Securities and Markets Authorities. (2017). *ESMA alerts investors to the high risks of initial coin offerings (ICOs).* https://www.esma.europa.eu/sites/default/files/library/esma50-157-829_ico_statement_investors.pdf

European Parliament and European Council. (2020). Regulation (EU) 2020/1503 of the European Parliament and of the Council of 7 October 2020 on European crowdfunding service providers for business, and amending Regulation (EU) 2017/1129 and Directive (EU)

2019/1937. https://eur-lex.europa.eu/legal-content/EN/TXT/PDF/?uri=CELEX:32020R1503&from=EN

Fisch, C. (2019). Initial coin offerings (ICOs) to finance new ventures. *Journal of Business Venturing, 34*(1), 1–22.

Fisch, C., & Momtaz, P. P. (2020). Institutional investors and post-ICO performance: An empirical analysis of investor returns in initial coin offerings (ICOs). *Journal of Corporate Finance, 64*, 101679.

Fisch, C., Meoli, M., & Vismara, S. (2022). Does blockchain technology democratize entrepreneurial finance? An empirical comparison of ICOs, venture capital, and REITs. *Economics of Innovation and New Technology, 31*(1–2), 70–89. https://doi.org/10.1080/10438599.2020.1843991

Fleming, L., & Sorenson, O. (2016). Financing by and for the masses: An introduction to the special issue on crowdfunding. *California Management Review, 58*(2), 5–19.

Frame, W. S., & White, L. J. (2014). *Technological change, financial innovation, and diffusion in banking*. In A. N. Berger, P. Molyneux, & J. O. S. Wilson (Eds.), *The Oxford Handbook of banking* (2nd ed., pp. 1–5). Oxford University Press.

Frame, W. S., Wall, L. D., & White, L. J. (2018). *Technological change and financial innovation in banking: Some implications for fintech* (Federal Reserve Bank of Atlanta, Working Paper Series). https://www.frbatlanta.org/-/media/documents/research/publications/wp/2018/11-technological-change-and-financial-innovation-in-banking-some-implications-for-fintech-2018-10-02.pdf

FSB. (2017). *Financial stability implications from FinTech*. Supervisory and Regulatory Issues that Merit Authorities' Attention, 27 June 2017. https://www.fsb.org/wp-content/uploads/R270617.pdf

Giudici, G., Milne, A., & Vinogradov, D. (2020). Cryptocurrencies: Market analysis and perspectives. *Journal of Industrial and Business Economics, 47*, 1–18.

Goldstein, I., Jiang, W., & Karolyi, G. A. (2019). To FinTech and beyond. *The Review of Financial Studies, 32*(5), 1647–1661.

Gopal, M., & Schnabl, P. (2022). The rise of finance companies and fintech lenders in small business lending. *The Review of Financial Studies, 35*(11), 4859–4901.

Hornuf, L., Kück, T., & Schwienbacher, A. (2022). Initial coin offerings, information disclosure, and fraud. *Small Business Economics, 58*, 1741–1759. https://doi.org/10.1007/s11187-021-00471-y

Howell, S. T., Niessner, M., & Yermack, D. (2020). Initial coin offerings: Financing growth with cryptocurrency token sales. *The Review of Financial Studies, 33*(9), 3925–3974.

KPMG. (2022, September). *Pulse of Fintech H1 2022*. https://assets.kpmg.com/content/dam/kpmg/xx/pdf/2022/08/pulse-of-fintech-h1-22.pdf

Lee, C. C., Li, X., Yu, C. H., & Zhao, J. (2021). Does fintech innovation improve bank efficiency? Evidence from China's banking industry. *International Review of Economics & Finance, 74*, 468–483.

Martino, P. (2019). Blockchain technology: Challenges and opportunities for banks. *International Journal of Financial Innovation in Banking, 2*(4), 314–333.

Martino, P. (2021). *Blockchain and banking: How technological innovations are shaping the banking industry*. Springer Nature.

Martino, P., Bell, G., Rasheed, A. A., & Bellavitis, C. (2022). Entrepreneurial finance and governance. In *Oxford research encyclopedia of business and management*. https://oxfordre.com/business/display/10.1093/acrefore/9780190224851.001.0001/acrefore-9780190224851-e-384;jsessionid=7FD107D9886E34B7F4C68560A5871F06?rskey=Mbmcxo&result=56

Martino, P., Bellavitis, C., & DaSilva, C. M. (2020a). Cryptocurrencies and entrepreneurial finance. In *The economics of cryptocurrencies* (pp. 51–56). Routledge.

Martino, P., Wang, K. J., Bellavitis, C., & DaSilva, C. M. (2020b). An introduction to blockchain, cryptocurrency and initial coin offerings. In *New frontiers in entrepreneurial finance research* (pp. 181–206). World Scientific.

Momtaz, P. P. (2021). Initial coin offerings, asymmetric information, and loyal CEOs. *Small Business Economics, 57*(2), 975–997.

Moritz, A., & Block, J. H. (2016). Crowdfunding: A literature review and research directions. In *Crowdfunding in Europe* (pp. 25–53). Springer.

Nakamoto, S. (2008). *Bitcoin: A peer-to-peer electronic cash system*. https://bitcoin.org/bitcoin.pdf

Philippon, T. (2016). *The fintech opportunity* (National Bureau of Economic Research Working Paper Series, No. w22476). https://www.nber.org/system/files/working_papers/w22476/w22476.pdf

PWC. (2020). *FinTech calls for fuel to exploit a great, maturing and increasing potential*. https://www.pwc.com/it/it/industries/fintech/docs/PwC-FinTech-2020.pdf

Rossi, A., Vanacker, T., & Vismara, S. (2021). Equity crowdfunding: New evidence from US and UK markets. *Review of Corporate Finance, 1*(3–4), 407–453.

Securities and Exchange Commission. (2021). *Crowdfunding for investors—Updated investor bulletin*. https://www.investor.gov/introduction-investing/general-resources/news-alerts/alerts-bulletins/investor-bulletins/updated-11

Stulz, R. M. (2019). Fintech, bigtech, and the future of banks. *Journal of Applied Corporate Finance, 31*(4), 86–97.

Tapscott, A., & Tapscott, D. (2017). How blockchain is changing finance. *Harvard Business Review, 1*(9), 2–5.

Thakor, A. V. (2020). Fintech and banking: What do we know? *Journal of Financial Intermediation, 41*, 100833.

Vismara, S. (2018). Information cascades among investors in equity crowdfunding. *Entrepreneurship Theory and Practice, 42*(3), 467–497.

Yeoh, P. (2017). Regulatory issues in blockchain technology. *Journal of Financial Regulation and Compliance, 25*(2), 196–208.

Yermack, D. (2015). Is bitcoin a real currency? An economic appraisal. In *Handbook of digital currency* (pp. 31–43). Academic Press.

Yermack, D. (2017). Corporate governance and blockchains. *Review of finance, 21*(1), 7–31.

Yermack, D., & Fingerhut, A. (2019, May). Blockchain technology's potential in the financial system. In *Proceedings of the 2019 Financial Market's Conference.*

Zhao, J., Li, X., Yu, C. H., Chen, S., & Lee, C. C. (2022). Riding the FinTech innovation wave: FinTech, patents and bank performance. *Journal of International Money and Finance, 122*, 102552.

CHAPTER 6

An Introduction to Sustainable Finance

Abstract This chapter introduces the phenomena of sustainable finance, providing a discussion about its main implications for the banking industry and financial markets. In particular, this chapter examines the role of banks in sustainable finance, highlighting the main recent regulatory developments in the industry, and how banks are responding to the increased demand for their commitment to sustainability issues. Moreover, it explores recent developments in financial markets, where a host of new financial instruments are available to meet Environmental, Social and Governance (ESG) concerns.

Keywords Sustainability · ESG · Sustainable finance · Banks · Financial markets

6.1 The Rise of Sustainable Finance

Today, the world is facing great social, environmental and economic challenges. A growing number of scientists has over the past decade issued warnings about the dangers of climate change, environmental degradation, social issues and other related problems (e.g., poverty, economic inequalities, and, more recently, mitigating pandemic threats) (Levy, 2021; Pizzi et al., 2021). It has become clear that economic activities

P. Ferretti and P. Martino, *Banking and Financial Markets*,
https://doi.org/10.1007/978-3-031-32562-5_6

are strictly linked to society and the environment, making it urgent to reconcile economic and social development on the one hand, and environmental protection on the other, before the abovementioned problems become irreversible.

Accordingly, improved sustainability of industrial activities has become a key topic of discussion among policy-makers and industrial decision-makers around the world (Scordato et al., 2018; Stoycheva et al., 2018). In 2015, for example, the "Transforming our World: the 2030 Agenda for Sustainable Development"—including its 17 Sustainable Development Goals (SDGs) and 169 targets (Fig. 6.1)—was adopted by Heads of State and Government at a special United Nations (UN) summit.

The Agenda is a commitment to eradicate poverty and achieve sustainable development by 2030 worldwide, ensuring that no one is left behind. A key feature of the Agenda is that the SDGs are global in nature and universally applicable, taking into account national realities, capacities and levels of development and specific challenges. This means that all countries have a shared responsibility to achieve the SDGs, and all have a meaningful role to play locally, nationally as well as on the global scale. Moreover, the 2030 Agenda integrates in a balanced manner the three

Fig. 6.1 The 17 sustainable development goals (*Source* United Nations)

dimensions of sustainable development—economic, social and environmental. It is also indivisible, in a sense that it must be implemented as a whole, in an integrated rather than a fragmented manner, recognizing that the different goals and targets are closely interlinked.

In addition, the Paris Agreement—the first-ever universal, legally binding global climate change agreement—was adopted the same year at the Paris climate conference (COP21). It sets out a global framework to avoid dangerous climate change by limiting global warming to well below 2 °C and pursuing efforts to limit it to 1.5 °C. It also aims to strengthen countries' ability to deal with the impacts of climate change and support them in their efforts.

The EU (and its Member States) has played a leading role in the process that led to the adoption of the 2030 Agenda for Sustainable Development and its 17 SDGs, and it is among the almost 190 Parties that signed the Paris Agreement. Moreover, among the several initiatives undertaken by the EU in this respect, it is worth mentioning the *European Green Deal* which is a package of policy initiatives, which aims to set the EU on the path to a green transition, with the ultimate goal of reaching climate neutrality by 2050.[1] On this path towards sustainability, understood in a broad sense, *sustainable finance* plays a key role in delivering on the policy objectives under the European Grean Deal as well as the EU's international commitments on climate and sustainability objectives.

According to the European Commission (EC), sustainable finance refers to the process of taking *environmental*, *social* and *governance* (ESG) considerations into account when making investment decisions in the financial sector, leading to more long-term investments in sustainable economic activities and projects. *Environmental* considerations include climate change mitigation and adaptation, as well as the environment more broadly, for instance the preservation of biodiversity, pollution prevention and the circular economy. *Social* considerations refer to issues of inequality, inclusiveness, labour relations, investments in human capital and communities, as well as human rights issues. The *governance* of public and private institutions—including management structures, employees relations and executive remuneration—plays a fundamental

[1] See, for more details: https://commission.europa.eu/strategy-and-policy/priorities-2019-2024/european-green-deal_en.

role in ensuring the inclusion of social and environmental considerations in the decision-making process.

The EU has been at the forefront of efforts to build a financial system that supports sustainable growth. To this end, the Commission has since 2018 been developing a comprehensive policy agenda on sustainable finance, comprising the "*action plan on financing sustainable growth and the development of a renewed sustainable finance strategy*" in the framework of the European Green Deal, and the new "*strategy for financing the transition to a sustainable economy*" on 6 July 2021. All these initiatives aim to support the financing of the transition to a sustainable economy by proposing actions to pursue.

In the next sections, we explore how the financial system can support the transition to a sustainable economy, focusing particularly on the role of the banking industry and financial markets, where a host of new instruments are available to finance sustainable investments.

6.2 Banking and Sustainable Finance

Given its role in allocating financial resources to the economy, the banking sector is called upon to play a leading role in a new model of economic development geared towards sustainability developments (Bose et al., 2018; Chen et al., 2022; Cosma et al., 2020; Wu & Shen, 2013; Zimmermann, 2019). This is due to the specific activities of credit and financial intermediation, such as screening, monitoring, enforcement, the possible consideration of environmental, social and governance factors in funding provision and investment activities (Cosma et al., 2020). In other words, banks can play a pivotal role in financing the transition to a greener and more sustainable economy, favouring—in the process of granting credit—those business initiatives oriented towards combining economic-financial objectives with social and environmental ones and, more generally, promoting sustainable investment initiatives in the allocation of savings (Miralles-Quirós et al., 2019; Umar et al., 2021). The underlying assumption is that if banks are willing to change their investment strategies encompassing ESG factors, then sustainable growth could be possible (Ahmed et al., 2018; La Torre et al., 2021).

The EU banking regulators have pictured a precise ESG path with the aim to further connect the banking industry with sustainability (EBA, 2019; ECB, 2020). For example, through the "Action plan on sustainable finance" (2019), the EBA has defined a roadmap providing for the regulation of the integration of ESG factors within the strategy, business model and risk management of banks by 2025, as well as the definition of prudential supervision requirements and sustainability disclosure criteria. Moreover, at the end of 2022, the EBA published a roadmap outlining the objectives and timeline for delivering mandates and tasks in the area of sustainable finance and ESG risks. This roadmap—which builds on and replaces the EBA's first action plan on sustainable finance—explains the EBA's sequenced and comprehensive approach over the next three years to integrate ESG risks considerations in the banking framework and support the EU's efforts to achieve the transition to a more sustainable economy. It ensures continuity of actions assumed under the previous action plan, while accommodating the necessary adjustments following the market and regulatory developments, including new mandates and new areas of focus, as reported in Fig. 6.2. Specifically, in the area of transparency and disclosures, the EBA will continue its work related to the development and implementation of institutions' ESG risks and wider sustainability disclosures. Similarly, the EBA will continue its efforts to ensure that ESG factors and risks are adequately integrated in institutions' risk management framework and in their supervision, including through further developments on climate stress tests. In the area of prudential regulation, the EBA has initiated an assessment of whether amendments to the existing prudential treatment of exposures to incorporate environmental and social considerations would be justified. Furthermore, the EBA will contribute to the development of green standards and labels and measures to address emerging risks in this field, such as greenwashing. Finally, the EBA will be assessing and monitoring developments in sustainable finance and institutions' ESG risk profile, including on the basis of the expected supervisory reporting.

Among the different initiatives undertaken by the EBA in this field, it is also worth mentioning the "*Guidelines on loan origination and monitoring*"[2] (2020), which introduced prominently environmentally sustainable lending dimensions, and set requirements for institutions to consider ESG factors and associated risks in their credit policies and procedures. This means that institutions should incorporate ESG factors and associated risks in their credit risk appetite and risk management policies, credit risk policies and procedures, adopting a holistic approach.

[2] This document represents the first specific policy measure developed by the EBA incorporating sustainability considerations.

Fig. 6.2 EBA's key objectives of its roadmap on sustainable finance (*Source* European Banking Authority [2022])

Following the increased demand to account for ESG issues, a growing number of banks has in recent years strengthened their commitment to sustainability objectives, considering ESG factors in the assessment of customers' lending and credit scoring, creating specific products and services to finance eco-sustainable investments or carrying out initiatives consistent with environmental and social objectives (e.g., Avrampou et al., 2019; Birindelli et al., 2015; Broccardo et al., 2016; Cosma et al., 2020). As an example, it is useful to recall the Principles for Responsible Banking, signed by 130 international banks (e.g., Citigroup, Deutsche Bank, Intesa Sanpaolo, Santander, Crédit Agricole, etc.), whose aim is to align products, services and business strategies not only with individual clients and investors' needs and expectations, but also with social values and objectives, in particular, with the objectives adopted with the Paris Agreement on climate change and the United Nations SDGs.

Banks' current sustainability offerings are typically incorporated in traditional lending products. For example, *green loans* have attracted worldwide attention in recent years due to the increasingly striking environmental problems caused by economic activities (Li et al., 2018). According to the Green Loans Principles (Box 6.1), green loans are any type of loan instrument made available exclusively to finance or re-finance, in whole or in part, new and/or existing eligible Green Projects. Hence, they are fundamentally loans meant for environmentally friendly purposes, such as reducing CO_2 emissions, or purposes contributing to the green transition in society. For example, households may obtain green loans to purchase an electric car, or for installing solar cells on the roof of a house, while companies may obtain green loans to build zero emission buildings.

Box 6.1: The Green Loan Principles (GLP)

The Green Loan Principles (GLP) have been developed by an experienced working party, consisting of representatives from leading financial institutions active in the global syndicated loan markets (i.e. Loan Market Association, Asia Pacific Loan Market Association, Loan Syndications & Trading Association), with a view to promoting the development and integrity of the green loan product. Specifically, the GLP comprise voluntary recommended guidelines, to be applied by market participants on a deal-by-deal basis depending on the underlying characteristics of the transaction, that seek to promote integrity in the development of the green loan market by clarifying the instances in which a loan may be categorized as "green". It is worth noting that these principles build on and refer to the Green Bond Principles (GBP) of the International Capital Market Association, which we address in the next paragraph.

To be qualified as a green loan, a loan should be structured in alignment with the Green Loan Principles, which provide an international standard based on the following four core components (The World Bank, 2021):

1. *Use of Proceeds*: Designated green projects should provide clear environmental benefits, which will be assessed, measured, and reported by the borrower.
2. *Process for Project Evaluation and Selection*: The borrower of a green loan should clearly communicate how it is organized to assess and select projects that will receive loan proceeds. In addition, the

borrower explains how it will manage environmental and social risk of eligible projects.

3. *Management of Proceeds*: The proceeds of a green loan should be credited to a dedicated account or tracked by the borrower to maintain transparency and promote the integrity of the product.
4. *Reporting*: The principles recommend the use of qualitative performance indicators and, where feasible, quantitative performance measures (e.g., energy capacity, electricity generation, greenhouse gas emissions reduced/avoided, etc.)

Other instruments in this area are *social loans*, which aim to facilitate and support economic activity that mitigates social issues and challenges and/or achieves positive social outcomes.

Finally, an innovative financing solution dedicated to the ESG transition of companies is *sustainability-linked loans* which feature a bonus mechanism that allows the company to obtain a discount on the cost of the loan upon achieving specific objectives key performance indicators (KPIs) in the ESG area (Box 6.2). In other words, they are characterized by interest margins that change depending on the borrower's ability to meet certain sustainability metrics established by the borrower and lender at the time of loan origination, thereby incentivizing companies' sustainability performance (Pohl et al., 2023). These KPIs can refer to all the three dimensions (ESG) of the sustainability or specific areas, such as climate change and diversity among others.

Box 6.2: An Example of Sustainability-Linked Loans

The Italian banking group Intesa Sanpaolo supported in 2022 the Italian shipbuilding company Fincantieri with a €500 million "sustainability linked" revolving credit facility for the construction of cruise ships. The three-year financing is characterized by a pricing mechanism linked to the achievement of two performance indicators ("KPIs"), i.e. *sustainability score*, assigned annually by S&P Global and *energy efficiency* of the cruise ships delivered each year, thus linking the financing to the achievement of precise sustainability targets.

It is worth noting that the working party, consisting of representatives from leading financial institutions active in the global syndicated loan markets, provides specific guidelines for each of these types of loans.

However, while principles for green and social loans are based around the abovementioned four core components (i.e. Use of Proceeds, Process for Project Evaluation and Selection, Management of Proceeds and Reporting), the principles for sustainability-linked loans are based around the following five core components: (1) Selection of KPIs, (2) Calibration of Sustainability Performance Targets (SPTs), (3) Loan Characteristics, (4) Reporting and (5) Verification.

Although banks have taken steps to account for ESG factors in their activities, as discussed before, empirical studies (BlackRock FMA, 2021; EBA, 2020; Ehlers et al., 2022; Yip & Bocken, 2018) show that there are still challenges to address, particularly with regard to incorporate sustainability within banks' decision-making processes and business functions, such as risk management. For example, several studies (Ehlers et al., 2022; Kleimeier & Viehs, 2018; Nguyen et al., 2022; Yip & Bocken, 2018) have investigated the pricing of climate-related risks in bank loan markets. From this perspective of analysis, it emerges that works are still in progress and further efforts by banks are needed. For example, a study by Yip and Bocken (2018) shows that although banks today have environment and social assessment guidelines in place for negative screening within the lending process; very few banks adjust the interest rate according to their borrowers' degrees of sustainability performance. Moreover, a recent study by Ehlers et al. (2022) regarding the pricing of carbon risk in the syndicated loan market shows that, while there is a risk premium charged to borrowing firms with higher carbon intensities since the Paris Agreement, the level of the premium appears small relative to the material risks. With specific regard to the EU context, it is worth mentioning a study made by BlackRock Financial Markets Advisory (2021) on 42 EU banks, which explores the integration of ESG factors into banks' risk management processes, business strategies and investment policies, as well as into prudential supervision (BlackRock FMA, 2021). This study shows that ESG integration in banks is at an early stage, and the pace of implementation needs to be accelerated in order to achieve effective ESG integration into banks' risk management and business strategies, as well as prudential supervision. Specifically, it reveals that despite the fact that the majority of interviewed banks have strategies in place for the integration of ESG into lending and investments; there are seldom comprehensive KPIs or processes in place to monitor their implementation at an in-depth level. Moreover, while some banks state that they tend to align their ESG strategy with international agreements (such as the UN SDGs or the

Paris Agreement), few banks have publicly specified concrete action plans to achieve those aims and disclose the progress they have made towards them. Finally, the integration of ESG risks into risk models, as well as stress testing, ICAAP, ILAAP and regulatory processes, are seen to be at a very early stage.

Hence, a gap still exists between what regulators ask and what is delivered in practice, which highlights the need for further efforts by banks to allow for a full integration of ESG factors within their business. Notably, existing literature suggests that banks that aspire to be sustainable should bear in mind that the sustainability must be an integral part of banks' DNA and of their overall strategy (e.g., EBA, 2019; Procopio et al., 2020; Galletta et al., 2021) and, therefore, integrated into their business plans, risk management, internal control framework and decision-making processes. This requires the creation and dissemination of a culture of "sustainability", starting with the bank's strategic supervisory and management bodies, then moving on to the functions involved in the operational management of credit processes, finance and services provided to customers, up to and including the risk management function.

6.3 Sustainable Finance: New Developments in Financial Markets

The financing of sustainable investments does not only concern banks, but it is a trend that also characterizes the most important institutional investors globally (e.g., Black Rock), which are assigning increasing importance to the issue of sustainability when evaluating investments (Cunha et al., 2021; Edmans & Kacperczyk, 2022; Gibson Brandon et al., 2022; Van Duuren et al., 2016; Widyawati, 2020). Nowadays, more and more investors indeed look to align their portfolios with their financial goals and internationally recognized sustainability goals such as Paris Agreement or SDGs by the UN. The prevailing evidence that sustainability profiles positively influence the market performance of investments has prompted investors to integrate ESG information into their investment decisions (Bello, 2005; Schröder, 2004; Statman, 2000).

Sustainable and responsible investments (SRIs) strategies are, in a broad sense, a long-term oriented investment approach which integrates ESG factors into investment decision-making, or, in other words, in the research, analysis and selection process of securities within an investment portfolio. Unlike conventional types of investments, SRIs apply a set of investment screens to select or exclude assets based on ecological, social, corporate governance or ethical criteria (Cunha et al., 2021; Renneboog et al., 2008; Riedl and Smeets, 2017; Sparkes and Cowton, 2004).

The growing demand for sustainable investments points out the need for capital markets to join the race of reaching the ambitious Paris Agreement and the UN 2030 Agenda for sustainable development (Popescu et al., 2021). In this changing context, *green bonds*—a growing subset of the ESG investing universe—have become increasingly popular among companies and investors looking to reorient capital flows towards more sustainable investments (Barua & Chiesa, 2019; Quirici, 2020; Sangiorgi & Schopohl, 2021). Green bonds are innovative financial instruments in which the proceeds are invested exclusively in green projects that generate climate or other environmental benefits (Carè, 2018; Flammer, 2021; Gianfrate & Peri, 2019). With a green bond, the issuer gets the capital to finance green projects, while the investors receive fixed income in the form of interest. At maturity, the principal is repaid, unless the issuer goes bankrupt. Hence, these instruments adopt a similar structure to other bonds, but have a specific constraint on the destination of the funds raised, for which an annual statement is required to inform investors of the actual use. The projects that can be financed cover a wide range of initiatives: energy efficiency, renewable energy, transport networks, natural resource management, among others. To qualify as a "certified" green bond, companies have to undergo third-party verification to establish that the proceeds are funding projects that generate environmental benefits. This means that these instruments have to meet precise conditions defined by international guidelines, including the principles of the ICMA (International Capital Market Association)[3] and the standards of the Climate Bond Initiative[4] (Ehlers & Packer, 2017). For example, the Green Bond Principles (GBP) by the ICMA are voluntary process guidelines that recommend transparency and disclosure and

[3] The *International Capital Market Association* represents financial institutions active in the international capital market worldwide. Its mission is to promote resilient well-functioning international and globally coherent cross-border debt securities markets, which are essential to fund sustainable economic growth and development.

[4] Climate Bonds Initiative is an international organization working to mobilize global capital for climate action.

promote integrity in the development of the green bond market by clarifying the approach for issuance of a green bond. Specifically, the Principles provide prospective issuers with guidance on the key components of green bond issuance, namely: (1) the use of proceeds for environmentally sustainable activities; (2) a process for determining project eligibility; (3) management of the proceeds in a transparent fashion that can be tracked and verified; and (4) annual reporting on the use of proceeds (ICMA, 2021a).

In addition to the abovementioned international certification mechanisms available to any issuer, it is worth noting that many jurisdictions around the world are developing their own taxonomies of what constitutes eligibility as a green bond. For example, the European Commission published in 2021 its proposal for a regulation to establish voluntary EU Green Bonds Standards (GBS) to improve the effectiveness, transparency, comparability and credibility of the market. This proposed Regulation sets a gold standard for how companies and public authorities can use green bonds to raise funds on capital markets to finance such ambitious large-scale investments, while meeting tough sustainability requirements and protecting investors. As the EC mentioned, this will be useful for both issuers and investors of green bonds: while issuers will have a robust tool to demonstrate that they are funding legitimate green projects aligned with the EU taxonomy, investors will be able to more easily assess, compare and trust that their investments are sustainable, thereby reducing the risks posed by greenwashing.

Since the European Investment Bank (EIB) pioneered the green bonds market by issuing the world's first Climate Awareness Bond (CAB) in late 2007, the European market has remarkably grown, with an average of 50% per year in the period 2015–2020. Currently, the EU is a global leader in green bonds, with 51% of the worldwide volume of green bonds being issued (Barua & Chiesa, 2019; Cicchiello et al., 2022). Moreover, green bonds have attracted a variety of issuer types (Sangiorgi & Schopohl, 2021) including sovereigns, municipalities, supranational organizations as well as financial and non-financial corporations, particularly those in the utility, automotive and real estate sectors.

In light of the growing interest in the world of ESG on the part of financial markets and institutions, there is now an evolution of sustainable financial products, designed according to standards and regulations issued at an international level. Alongside green bonds, other instruments include *social bonds*, the proceeds of which are earmarked for projects with a positive social impact, as well as *sustainability bonds*, whose funds are tied to projects with both environmental and social profiles. Specifically, social bonds are the use of proceeds bonds that raise funds for new and existing projects that address or mitigate a specific social issue and/or seek to achieve positive social outcomes. Examples of social project categories include providing and/or promoting: affordable basic infrastructure (e.g., clean drinking water, sewers, transport), access to essential services (e.g., health, education and vocational training, healthcare, financing and financial services), affordable housing, food security, etc. Sustainability bonds, on the other hand, are intended to finance green and social initiatives including, for example, energy efficiency and renewable energy, in addition to water efficiency, social infrastructure, etc.

Finally, since 2019 there has been a strong increase in the issuance of a new type of sustainable finance instruments, i.e. *sustainability-linked bonds* (Antilici et al., 2022). According to the Principles of the ICMA, sustainability-linked bonds (SLBs) are any type of bond instrument for which the financial and/or structural characteristics can vary depending on whether the issuer achieves predefined sustainability/ESG objectives. In other words, penalties or benefits are provided to the issuer in the event of failure to achieve the predefined objectives or to meet them. In this way, issuers are thus committing explicitly (including in the bond documentation) to future improvements in sustainability outcomes within a predefined timeline. Those objectives are measured through predefined KPIs and assessed against predefined Sustainability Performance Targets (SPTs). Some examples of KPIs, mainly used by non-financial companies, are the level of greenhouse gas emissions and the level of efficiency in the use of natural resources, such as the increased use of renewable energy sources. The Box 6.3 provides an example of sustainability-linked bonds.

Box 6.3. An Example of Sustainability-Linked Bonds

It is worth mentioning the first (multi-tranche) SLBs issue launched on the European market by the energy group Enel.

On 10 October 2019, Enel Finance International N.V., the Dutch-registered finance subsidiary of Enel SpA, launched a multi-tranche "sustainable" bond for institutional investors on the European market totalling 2.5 billion euros. The bond was linked to the achievement of the United Nations Sustainable Development Goals (SDGs). In particular, this bond issue was linked to Enel's ability to achieve the following Sustainable Development Goals:

i. SDG 7 "Affordable and clean energy", through the achievement, by 31 December 2021 of a percentage of installed renewable generation capacity (on a consolidated basis) equal to or greater than 55% of total consolidated installed capacity.
ii. SDG 13 "Climate action", through the achievement of a level of greenhouse gas emissions by 2030 equal to or less than 125 g of CO_2 per kWh, in line with the commitment to reduce Enel's direct greenhouse gas emissions per kWh by 70% by 2030 compared to the 2017 values, as certified by the Science Based Targets initiative (SBTi) and consistent with the Paris Agreement on climate change.

The issue was structured in several tranches. In order to explain the functioning of this type of bond, let us consider the one with the longest maturity, i.e. 500 million euros at a fixed rate of 1.125%, maturing 17 October 2034 and linked to SDG 13 "Climate action". The specific conditions were: if the sustainable target indicated under point (ii) above as of 31 December 2030 is achieved, the interest rate (set at the time of issue, i.e. 1.125%) will remain unchanged to maturity; conversely, if that target is not achieved, a step-up mechanism will be applied, increasing the rate by 25 bps as of the first interest period subsequent to the publication of the report issued by a third-party expert charged with validating the methodology for measuring CO_2 emissions applied by the Group.

It is worth mentioning that the ICMA provides specific guidelines for the issue of each of these types of bonds. However, while principles for green, social and sustainability bonds are based around the abovementioned four core components (i.e. Use of Proceeds, Process for Project Evaluation and Selection, Management of Proceeds and Reporting), the principles for sustainability-linked bonds are based around the following five core components: (1) Selection of Key Performance Indicators (KPIs), (2) Calibration of Sustainability Performance Targets (SPTs), (3) Bond characteristics, (4) Reporting and (5) Verification (ICMA, 2021a, 2021b, 2021c, 2021d).

Compared to other ESG instruments, SLBs are more versatile, as they can be adapted to the sustainability strategy of the issuer and are unconstrained in the allocation of proceeds, which indeed is not reserved for specific purposes. The resulting greater freedom in the use of funds for the issuer represents the main difference from green bonds, social bonds and sustainability bonds. Indeed, as aforementioned, the latter state that the proceeds from placements are exclusively used to finance specific environmental and/or social projects defined at issuance. Accordingly, while for these categories of instruments the relationship between financing and sustainability is ensured by limitations on the use of funds (compliance with which is periodically verified), the link between the raising of funds through the issuance of SLBs and the sustainability of their use is substantiated: (i) in the credibility of the incentives for the issuer that derive from the penalty or premium mechanisms to be applied according to the degree of achievement of the sustainability performance defined at issuance and (ii) in the periodic verification of such performance.

To conclude, it is worth noting that, at present, there is no evidence of potential displacement effects of SLBs emissions on the other ESG bonds emissions (or vice versa). By contrast, there are several examples of issuers that, as part of their overall sustainability strategy, have issued various types of ESG bond instruments, receiving a positive response from investors.

References

Ahmed, S. U., Ahmed, S. P., & Hasan, I. (2018). Why banks should consider ESG risk factors in bank lending? *Banks & Bank Systems, 13*(3), 71–80.

Antilici P., Mosconi G., & Russo L. (2022). *Quando innovazione finanziaria e finanza sostenibile si incontrano: i Sustainability-Linked Bonds*. Bank of Italy. https://www.bancaditalia.it/pubblicazioni/mercati-infrastrutture-e-sistemi-di-pagamento/questioni-istituzionali/2022-022/N.22-MISP.pdf.

Avrampou, A., Skouloudis, A., Iliopoulos, G., & Khan, N. (2019). Advancing the sustainable development goals: Evidence from leading European banks. *Sustainable Development, 27*(4), 743–757.

Barua, S., & Chiesa, M. (2019). Sustainable financing practices through green bonds: What affects the funding size? *Business Strategy and the Environment, 28*(6), 1131–1147.

Bello, Z. Y. (2005). Socially responsible investing and portfolio diversification. *Journal of Financial Research, 28*(1), 41–57.

Birindelli, G., Ferretti, P., Intonti, M., & Iannuzzi, A. P. (2015). On the drivers of corporate social responsibility in banks: Evidence from an ethical rating model. *Journal of Management & Governance, 19*, 303–340.

BlackRock Financial Markets Advisory. (2021). *Development of tools and mechanisms for the integration of ESG factors into the EU banking prudential framework and into banks' business strategies and investment policies*. https://op.europa.eu/en/publication-detail/-/publication/ce43e64f-06e0-11ec-b5d3-01aa75ed71a1/language-en.

Bose, S., Khan, H. Z., Rashid, A., & Islam, S. (2018). What drives green banking disclosure? An institutional and corporate governance perspective. *Asia Pacific Journal of Management, 35*, 501–527.

Broccardo, E., Costa, E., & Mazzuca, M. (2016). Corporate social responsibility and financial performance in Italian co-operative banks. In *Routledge handbook of social and sustainable finance* (pp. 280–295). Routledge.

Carè, R., & Carè, R. (2018). Exploring the role of banks in sustainable development. *Sustainable Banking: Issues and Challenges*, 39–64.

Chen, Z., Mirza, N., Huang, L., & Umar, M. (2022). Green banking—Can financial institutions support green recovery? *Economic Analysis and Policy, 75*, 389–395.

Cicchiello, A. F., Cotugno, M., Monferrà, S., & Perdichizzi, S. (2022). Which are the factors influencing green bonds issuance? Evidence from the European bonds market. *Finance Research Letters, 50*, 103190.

Cosma, S., Venturelli, A., Schwizer, P., & Boscia, V. (2020). Sustainable development and European banks: A non-financial disclosure analysis. *Sustainability, 12*(15), 6146.

Cunha, F. A. F. D. S., Meira, E., & Orsato, R. J. (2021). Sustainable finance and investment: Review and research agenda. *Business Strategy and the Environment, 30*(8), 3821–3838.

Edmans, A., & Kacperczyk, M. (2022). Sustainable finance. *Review of Finance, 26*(6), 1309–1313.

Ehlers, T., & Packer, F. (2017). Green bond finance and certification. *BIS Quarterly Review September*.

Ehlers, T., Packer, F., & de Greiff, K. (2022). The pricing of carbon risk in syndicated loans: Which risks are priced and why? *Journal of Banking & Finance, 136*, 106180.

European Banking Authority—EBA. (2019). *EBA action plan on sustainable finance*. https://www.eba.europa.eu/sites/default/documents/files/document_library/EBA%20Action%20plan%20on%20sustainable%20finance.pdf.

European Banking Authority—EBA. (2021). *Report on management and supervision of ESG risks for credit institutions and investment firms*. https://www.eba.europa.eu/sites/default/documents/files/document_library/Publications/Reports/2021/1015656/EBA%20Report%20on%20ESG%20risks%20management%20and%20supervision.pdf.

European Banking Authority. (2020). *Final report guidelines on loan origination and monitoring*. https://www.eba.europa.eu/sites/default/documents/files/document_library/Publications/Guidelines/2020/Guidelines%20on%20loan%20origination%20and%20monitoring/884283/EBA%20GL%202020%2006%20Final%20Report%20on%20GL%20on%20loan%20origination%20and%20monitoring.pdf

European Banking Authority. (2022). *Roadmap on sustainable finance*. https://www.eba.europa.eu/sites/default/documents/files/document_library/Publications/Reports/2022/ESG%20roadmap/1045378/EBA%20Roadmap%20on%20Sustainable%20Finance.pdf

European Central Bank. (2020). *Guide on climate-related and environmental risks supervisory expectations relating to risk management and disclosure*. https://www.bankingsupervision.europa.eu/ecb/pub/pdf/ssm.202011finalguideonclimate-relatedandenvironmentalrisks~58213f6564.en.pdf

European Commission—EC. (2018). *Action plan: Financing sustainable growth*. https://eur-lex.europa.eu/legal-content/EN/TXT/HTML/?uri=CELEX:52018DC0097&from=EN

Flammer, C. (2021). Corporate green bonds. *Journal of Financial Economics, 142*(2), 499–516.

Galletta, S., Mazzù, S., & Naciti, V. (2021). Banks' business strategy and environmental effectiveness: The monitoring role of the board of directors and the managerial incentives. *Business Strategy and the Environment, 30*(5), 2656–2670.

Gianfrate, G., & Peri, M. (2019). The green advantage: Exploring the convenience of issuing green bonds. *Journal of Cleaner Production, 219*, 127–135.

Gibson Brandon, R., Glossner, S., Krueger, P., Matos, P., & Steffen, T. (2022). Do responsible investors invest responsibly? *Review of Finance, 26*(6), 1389–1432.

International Market Capital Association. (2021a). *Green bond principles-voluntary process guidelines for issuing green bonds*. ICMA, Paris. https://www.icmagroup.org/assets/documents/Sustainable-finance/2022-updates/Green-Bond-Principles_June-2022-280622.pdf

International Market Capital Association. (2021b). *Social bond principles-voluntary process guidelines for issuing social bonds*. ICMA, Paris. https://www.icmagroup.org/assets/documents/Sustainable-finance/2022-updates/Social-Bond-Principles_June-2022v3-020822.pdf

International Market Capital Association. (2021c). *Sustainability bond guidelines*. ICMA, Paris. https://www.icmagroup.org/assets/documents/Sustainable-finance/2021-updates/Sustainability-Bond-Guidelines-June-2021-140621.pdf

International Market Capital Association. (2021d). *Sustainability-linked bond principles-voluntary process guidelines*. ICMA, Paris. https://www.icmagroup.org/assets/documents/Regulatory/Green-Bonds/June-2020/Sustainability-Linked-Bond-Principles-June-2020-171120.pdf

Kleimeier, S., & Viehs, M. (2018, January 7). Carbon disclosure, emission levels, and the cost of debt. *Emission Levels, and the Cost of Debt*.

La Torre, M., Leo, S., & Panetta, I. C. (2021). Banks and environmental, social and governance drivers: Follow the market or the authorities? *Corporate Social Responsibility and Environmental Management, 28*(6), 1620–1634.

Levy, D. L. (2021). COVID-19 and global governance. *Journal of Management Studies, 58*(2), 562.

Li, Z., Liao, G., Wang, Z., & Huang, Z. (2018). Green loan and subsidy for promoting clean production innovation. *Journal of Cleaner Production, 187*, 421–431.

Loan Market Association, Asia Pacific Loan Market Association, Loan Syndications & Trading Association. All rights reserved. (2021a). *Green loan principles supporting environmentally sustainable economic activity*. UK. https://www.lsta.org/content/green-loan-principles/

Loan Market Association, Asia Pacific Loan Market Association, Loan Syndications & Trading Association. All rights reserved. (2021b). *Social loans principles*. UK. Available at: https://www.lsta.org/content/social-loan-principles-slp/

Loan Market Association, Asia Pacific Loan Market Association, Loan Syndications & Trading Association. All rights reserved. (2022). *Sustainability-linked loan principles supporting environmentally and socially sustainable economic activity*. UK. https://www.lsta.org/content/sustainability-linked-loan-principles-sllp/

Miralles-Quirós, M. M., Miralles-Quirós, J. L., & Redondo-Hernández, J. (2019). The impact of environmental, social, and governance performance on stock prices: Evidence from the banking industry. *Corporate Social Responsibility and Environmental Management, 26*(6), 1446–1456.

Nguyen, D. D., Ongena, S., Qi, S., & Sila, V. (2022). Climate change risk and the cost of mortgage credit. *Review of Finance, 26*(6), 1509–1549.

Pizzi, S., Rosati, F., & Venturelli, A. (2021). The determinants of business contribution to the 2030 Agenda: Introducing the SDG Reporting Score. *Business Strategy and the Environment, 30*(1), 404–421.

Pohl, C., Schüler, G., & Schiereck, D. (2023). Borrower-and lender-specific determinants in the pricing of sustainability-linked loans. *Journal of Cleaner Production, 385*, 135652.

Popescu, I. S., Hitaj, C., & Benetto, E. (2021). Measuring the sustainability of investment funds: A critical review of methods and frameworks in sustainable finance. *Journal of Cleaner Production, 314*, 128016.

Procopio, G., Trotta, A., Strano, E., & Iannuzzi, A. P. (2020). Green banking in Italy: Current and future challenges. *Contemporary Issues in Sustainable Finance: Creating an Efficient Market Through Innovative Policies and Instruments*, 205–258.

Quirici, M. C. (2020). The increasing importance of green bonds as instruments of impact investing: Towards a New European Standardisation. *Contemporary Issues in Sustainable Finance: Creating an Efficient Market Through Innovative Policies and Instruments*, 177–203.

Renneboog, L., Ter Horst, J., & Zhang, C. (2008). Socially responsible investments: Institutional aspects, performance, and investor behavior. *Journal of Banking & Finance, 32*(9), 1723–1742.

Riedl, A., & Smeets, P. (2017). Why do investors hold socially responsible mutual funds? *The Journal of Finance, 72*(6), 2505–2550.

Sangiorgi, I., & Schopohl, L. (2021). Why do institutional investors buy green bonds: Evidence from a survey of European asset managers. *International Review of Financial Analysis, 75*, 101738.

Schröder, M. (2004). The performance of socially responsible investments: Investment funds and indices. *Financial Markets and Portfolio Management, 18*(2), 122.

Scordato, L., Klitkou, A., Tartiu, V. E., & Coenen, L. (2018). Policy mixes for the sustainability transition of the pulp and paper industry in Sweden. *Journal of Cleaner Production, 183*, 1216–1227.

Sparkes, R., & Cowton, C. J. (2004). The maturing of socially responsible investment: A review of the developing link with corporate social responsibility. *Journal of Business Ethics, 52*, 45–57.

Statman, M. (2000). Socially responsible mutual funds (corrected). *Financial Analysts Journal, 56*(3), 30–39.

Stoycheva, S., Marchese, D., Paul, C., Padoan, S., Juhmani, A. S., & Linkov, I. (2018). Multi-criteria decision analysis framework for sustainable manufacturing in automotive industry. *Journal of Cleaner Production, 187*, 257–272.

The World Bank. (2021). *What you need to know about green loans*. https://www.worldbank.org/en/news/feature/2021/10/04/what-you-need-to-know-about-green-loans

Umar, M., Ji, X., Mirza, N., & Naqvi, B. (2021). Carbon neutrality, bank lending, and credit risk: Evidence from the Eurozone. *Journal of Environmental Management, 296*, 113156.

Van Duuren, E., Plantinga, A., & Scholtens, B. (2016). ESG integration and the investment management process: Fundamental investing reinvented. *Journal of Business Ethics, 138*, 525–533.

Widyawati, L. (2020). A systematic literature review of socially responsible investment and environmental social governance metrics. *Business Strategy and the Environment, 29*(2), 619–637.

Wu, M. W., & Shen, C. H. (2013). Corporate social responsibility in the banking industry: Motives and financial performance. *Journal of Banking & Finance, 37*(9), 3529–3547.

Yip, A. W., & Bocken, N. M. (2018). Sustainable business model archetypes for the banking industry. *Journal of Cleaner Production, 174*, 150–169.

Zimmermann, S. (2019). Same same but different: How and why banks approach sustainability. *Sustainability, 11*(8), 2267.

Index

P. Ferretti and P. Martino, *Banking and Financial Markets*,
https://doi.org/10.1007/978-3-031-32562-5

GPSR Compliance
The European Union's (EU) General Product Safety Regulation (GPSR) is a set of rules that requires consumer products to be safe and our obligations to ensure this.

If you have any concerns about our products, you can contact us on

ProductSafety@springernature.com

In case Publisher is established outside the EU, the EU authorized representative is:

Springer Nature Customer Service Center GmbH
Europaplatz 3
69115 Heidelberg, Germany

www.ingramcontent.com/pod-product-compliance
Ingram Content Group UK Ltd.
Pitfield, Milton Keynes, MK11 3LW, UK
UKHW021022081026
15470UKWH00034B/1396
9783031325618